T0211908

SpringerBriefs in Physics

SpringerBriefs in Physics are a series of slim high-quality publications encompassing the entire spectrum of physics. Manuscripts for SpringerBriefs in Physics will be evaluated by Springer and by members of the Editorial Board. Proposals and other communication should be sent to your Publishing Editors at Springer.

Featuring compact volumes of 50 to 125 pages (approximately 20,000–45,000 words), Briefs are shorter than a conventional book but longer than a journal article. Thus, Briefs serve as timely, concise tools for students, researchers, and professionals.

Typical texts for publication might include:

- A snapshot review of the current state of a hot or emerging field
- A concise introduction to core concepts that students must understand in order to make independent contributions
- An extended research report giving more details and discussion than is possible in a conventional journal article
- A manual describing underlying principles and best practices for an experimental technique
- An essay exploring new ideas within physics, related philosophical issues, or broader topics such as science and society

Briefs allow authors to present their ideas and readers to absorb them with minimal time investment. Briefs will be published as part of Springer's eBook collection, with millions of users worldwide. In addition, they will be available, just like other books, for individual print and electronic purchase. Briefs are characterized by fast, global electronic dissemination, straightforward publishing agreements, easy-to-use manuscript preparation and formatting guidelines, and expedited production schedules. We aim for publication 8–12 weeks after acceptance.

More information about this series at https://link.springer.com/bookseries/8902

Parmod Kumar · Jitendra Pal Singh ·
Vinod Kumar · K. Asokan

Ion Beam Induced Defects and Their Effects in Oxide Materials

 Springer

Parmod Kumar 🆔
Department of Physics
J. C. Bose University of Science
and Technology, YMCA
Faridabad, Haryana, India

Vinod Kumar 🆔
Department of Physics
College of Natural and Computation
Science
Dambi Dollo University
Oromia Region, Ethiopia

Jitendra Pal Singh 🆔
Pohang Accelerator Laboratory
Pohang University of Science
and Technology
Pohang, Democratic People's Republic of
Korea

K. Asokan 🆔
Material Science Division
Inter University Accelerator Centre (IUAC)
New Delhi, India

Department of Physics
University of Petroleum and Energy Studies
(UPES)
Dehradun, India

ISSN 2191-5423   ISSN 2191-5431   (electronic)
SpringerBriefs in Physics
ISBN 978-3-030-93861-1   ISBN 978-3-030-93862-8   (eBook)
https://doi.org/10.1007/978-3-030-93862-8

This Springer imprint is published by the registered company Springer Nature Switzerland AG
The registered company address is: Gewerbestrasse 11, 6330 Cham, Switzerland

Acknowledgements

The authors gratefully acknowledge Inter-University Accelerator Centre (IUAC), New Delhi, India for all the support and encouragement. Dr. Parmod Kumar is thankful to the Department of Science and Technology (DST), Ministry of Science and Technology, New Delhi, India for support through the DST-Inspire Faculty Award [DST/INSPIRE/04/2015/003149]. Dr. Parmod Kumar would also like to express his gratitude towards Hon'ble Vice-Chancellor Prof. Dinesh Kumar, J. C. Bose University of Science and Technology, YMCA, Faridabad for their continuous encouragement during the book. The authors also thank Prof. Pablo Esquinazi, Division of Superconductivity and Magnetism, University of Leipzig, Germany for their guidance and support during the initial stage of this book.

Contents

About the Authors

Dr. Parmod Kumar is working as Assistant Professor at J. C. Bose University of Science and Technology, YMCA Faridabad, India. He has been a recipient of the prestigious DST-INSPIRE Faculty Award, DAAD Fellow, etc. His research interests include defect-induced magnetism in oxides, wastewater treatment, and irradiation/implantation-induced changes in oxides. He has published ~ 60 SCI-indexed publications.

Dr. Jitendra Pal Singh is working at Pohang Accelerator Laboratory, Pohang, South Korea. His research interests are irradiation studies in nanoferrites, thin films, and magnetic multilayers. He also studied the irradiation and implantation effects in ferrite thin films and nanoparticles. He has more than 100 SCI-indexed publications.

Dr. Vinod Kumar is working as Associate Professor, Department of Physics, College of Natural and Computation Science, Dambi Dollo University, Oromia Region, Ethiopia. Before that, he also worked at IIT Delhi, IUAC, New Delhi, India; UFS, Bloemfontein, South Africa; ICCF, UCA, France, and the University of Tulsa, USA. His research interest is oxide-based nanomaterials for solar cells as well as solid-state lighting application and the role of irradiation on it. He has published more than 120 SCI-indexed articles.

Dr. K. Asokan has served as a senior scientist in the Inter-University Accelerator Centre, New Delhi. He is currently working as a Professor in the Department of Physics and Centre for Interdisciplinary Research, University of Petroleum and Energy Studies (UPES) Dehradun, Uttarakhand. His research interests are in (i) understanding the ion beam interaction in materials and (ii) synchrotron-based spectroscopic studies. He has published over 400 papers in peer-reviewed journals with more than 7500 citations. He has collaborations with the research groups in Taiwan, Singapore, South Korea and Japan apart from various Indian universities.

Chapter 1
Introduction of Ion Beam Techniques

1.1 Introduction

Ion beam-based techniques have contributed significantly towards the prerequisites of society. These techniques are proven to be vital in the synthesis of materials, device fabrication as well as the characterization of the materials [1, 2]. Apart from the synthesis and characterization of materials, the manipulation of the physical properties and device characteristics are other important aspects. Ion beam-based sputtering methods are being extensively used to grow thin films of different kinds of materials [3–5]. Recently, the focussed ion beam (FIB) technique is used to prepare samples for various experimental techniques like transmission electron microscopy [6], X-ray imaging [7], etc. This technique is being further extended to microsystem technology [8].

Ion beam methods have shown the potential to utilize them for material synthesis [9]. Silicon-based materials [10, 11], graphene [12], nanowires [13], plasmonic nanostructures [14], and quantum dots [15] are some of the examples. Additionally, ion beam techniques play an important role in modifying the physical characteristics of the material [16]. Ion beam irradiation [17] and implantation [18] are two common and well-known techniques that are proven to be assets for this purpose. These techniques have found immense importance in recent years especially in semiconductor-based electronics. This book briefly discusses the ion beam processes and the potential applications in oxide semiconductors. Since the interaction of ions with materials plays an important role [19, 20], a brief description of this aspect is given in the next section.

© The Author(s), under exclusive license to Springer Nature Switzerland AG 2022
P. Kumar et al., *Ion Beam Induced Defects and Their Effects in Oxide Materials*,
SpringerBriefs in Physics,
https://doi.org/10.1007/978-3-030-93862-8_1

Table 1.1 Effect of ions of various energies on the solid surface

S. no.	Ion energy	Effect on solid surface	Reference
1.	10–100 eV/amu	Epitaxial layers	[21]
2.	1 keV/amu	Sputtering process	[22]
3.	100–300 keV/amu	Ion implantation	[23]
4.	~MeV/amu	Ion irradiation	[24]

1.1.1 Interaction of Ions with Matter

Ions incident on a solid influences its properties and are used to tailor the physical properties in different ways as shown in Table 1.1. Ions having energies of the order of keV/amu are commonly suitable for implantation purposes i.e. the ions beam resides inside the material due to less energy, however, the ions having the energy of the order of MeV/amu penetrate through the material due to high energy and utilized for irradiation purposes. Therefore, ion beam implantation and ion irradiation are two common approaches used for material modifications, and these techniques are discussed in detail in the subsequent chapters.

The interaction of an energetic ion depends upon its kinetic energy and the distance of the closest approach. This distance of the closest approach is related to the impact parameter (denoted by b) and is shown in Fig. 1.1 [25–27]. Based on the impact parameter (b) and atomic radius (a), the ion-matter interaction is classified as per the type of collisions.

If the impact factor is much larger than that of the atomic radius (*i.e.* b >> a), then the collisions are termed soft collisions. In this process, atoms can be excited to higher energy levels or ionized by the ejection of valence electrons and receive small energy. If b ~ a, the charged particle interacts with a single atomic electron, and

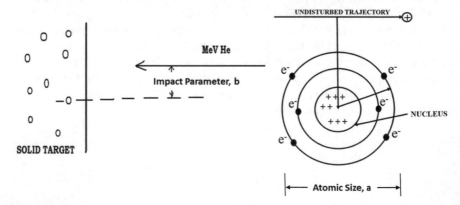

Fig. 1.1 Representation of impact parameter (denoted by b) during the interaction of energetic ion (4 MeV He) with solid material

this interaction results in energy loss along the track. Since the atomic spacing in a typical crystal is of the order of a few angstroms, the largest impact parameter is of a similar order. At this distance, the ion transfers a small amount of energy, $\approx 10\,\text{eV}$, to the valence electrons. The cross-section (σ) for this process is of the order of $\pi b^2 \approx 10^{-15}\,\text{cm}^2$, and it is responsible for the gradual energy loss of the ion as it penetrates the solid. For the smaller impact parameter ($b < a$), the interaction is with the inner electrons and correspondingly larger energy transfer occurs. For impact parameters comparable to the nuclear size, larger energy transfers occur ($\approx 100\,\text{keV}$); in billiard ball-like collisions resulting in the large angle of scattering of the incident ions [28].

1.2   Energy Loss in the Material/Target

As an ion penetrates in a solid, it continuously losses energy by collisions with the nuclei and electrons of the target atoms until it reaches thermal energies. A typical time scale for an ion to achieve the rest position is $\sim 10^{-14}$ s. The energy loss in the medium is a statistical energy transfer process that depends on the ion energy, type of ion and the physical properties of the target material. It is the energy loss experienced by an ion that ultimately determines the range to which it penetrates a target. The range, which is travelled by ions before coming to rest, is known as projected range, R. The energy loss per unit length is called stopping power and is denoted as 'S'. This loss is related to the reaction cross-section by the following relation

$$S = \frac{dE}{dx} = N \int T\sigma(E, T) \cdot dT,$$

where, N is the number density and T is energy transferred from ions having energy E to the material for a cross-section of σ. Energy loss in the medium is governed chiefly by the following processes-(1) Inelastic scattering with the electrons in the target (electronic energy loss, EEL) represented by $(dE/dx)_e$ or S_e and (2) elastic collision with the nucleus of the target atoms (nuclear energy loss, NEL) represented by $(dE/dx)_n$ or S_n [29].

The energy loss in the medium causes radiation damage (Fig. 1.2a). At low S_e values, cascades of atomic collisions lead to the formation of point-like defects (Fig. 1.2b). If S_e is sufficiently high, continuous cylindrical tracks (latent tracks) or columnar defects are produced in the ion's path (Fig. 1.2c). For moderate values of S_e, both kind of effects occur in the material (Fig. 1.2d). Thus, radiation damage depends on the amount of energy lost in the medium which depends on the initial energy or velocity of incident ions [30]. Figure 1.2e shows the stopping power (S) at various ion velocities (v).

In terms of ion velocity (v), elastic collisions are important if $v \ll v_0$, where v_0 is the orbital velocity of the electrons in the target. As the ion energy (E) increases, S_n diminishes as 1/E. Electronic interactions play an important role when the ion velocity

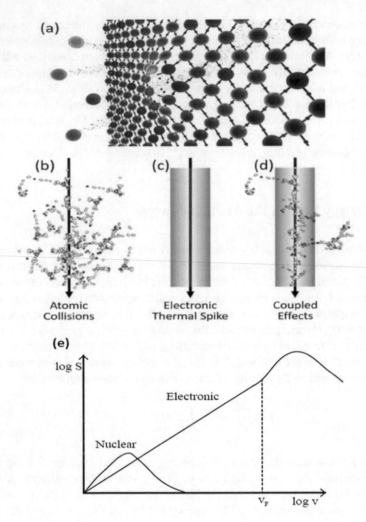

Fig. 1.2 a Interaction of ions with the target, **b** atomic collisions, **c** electronic thermal spike **d** coupled effects [30] and **e** Stopping power S versus ion velocity v. Here, V_F is the Fermi velocity. At higher velocity, electronic stopping is dominant compared to nuclear stopping

becomes comparable to the velocities of orbital electrons in the target materials. As a result, electronic energy loss (S_e) varies linearly in the velocity range $0.1v_0$ to $Z_1^{2/3}v_0$, Z_1 is the atomic number of target atoms [31]. Outside this range, it has lower values. The variation of S_e and S_n as functions of ion energy E is shown in Fig. 1.2. It is also pertinent to note that the relativistic effects must be taken into account for ion energies above 10 MeV/amu [31]. Thus, the various processes that occur in the material can be classified based on the amount of energy transfer to the lattice. These processes have a direct correlation with the impact parameters (Table 1.2).

Table 1.2 Classification of the types of collision and their corresponding energy transfer in terms of impact parameter for the process. S_e refers to energy transfer to electrons and S_n is energy transfer to the nuclei. Impact parameters and involved process during the interaction of energetic ions with the material

Impact parameter (b)		Energy transfer
~1 Å	Inelastic excitation of valence electrons	S_e~10 eV
~10^{-1} Å	Inelastic excitation of L-shell electrons	S_e~100 eV
~10^{-2} Å	Inelastic excitation of K-shell electrons	S_e~1 keV
~10^{-4} Å	Elastic scattering from Nuclei	S_n~100 keV

A large number of atoms in a target are set into motion by the electronic excitation in the wake of swift heavy ions (SHI) with kinetic energy (K.E.) $\geq$ 1 MeV during an irradiation experiment [32, 33]. The atomic mobility in the bulk leads to some well-known effects, such as anisotropic growth, change in electrical, magnetic, and optical properties. Although this area is now more than three decades old, there is no consensus about the physical processes responsible for all these observed effects.

1.3 The SRIM Computer Code

The defects produced by radiation damage lead to some interesting, macroscopically observable effects. The presence of these defects is chiefly determined by the amount of energy lost in the medium. Thus, the calculated parameters such as stopping values and projected range are required. SRIM (Stopping and Range of Ions in Matter) is a set of programs that calculate the stopping and range of ions (10 eV–2 GeV) into matter using a full quantum mechanical treatment of incident ion and target atoms collisions. This program is based on the Monte Carlo Simulation method developed by Ziegler and Biersack around 1983. This program is being continuously advanced with the key changes approximately after every five years. During the collisions, the incident ion and target atoms have a screened Coulomb collision, including exchange and correlation interactions between the overlapping electron shells. The incident ion has long-range interactions with the target, creating electron excitations and plasmon. These are described by including the collective electronic structure of the target and the interatomic bond structure when the calculation is set up. The charge state of the ion within the target is described using the concept of effective charge, which includes a velocity-dependent charge state and long-range screening due to the collective electronic sea of the target. This computer code helps one to determine the stopping energies, projected range, and energy straggling. This software is freely downloadable and available from the website: www.srim.org [34].

References

1. W. Li, X. Zhan, X. Song, S. Si, R. Chen, J. Liu, Z. Wang, J. He, X. Xiao, A review of recent applications of ion beam techniques on nanomaterial surface modification: design of nanostructures and energy harvesting. Small **15**, 1901820 (2019)
2. B. Schmidt, K. Wetzig, in *Ion Beams in Materials Processing and Analysis* (Springer, Wien, 2012)
3. C. Bundesmann, H. Neumann, The systematics of ion beam sputtering for deposition of thin films with tailored properties. J. Appl. Phys. **124**, 231102 (2018)
4. J.E.E. Baglin, Thin-film bonding using ion beam techniques-a review. IBM J. R and D **38**, 413–422 (1994)
5. M. Beckera, M. Gies, A. Polity, S. Chatterjee, P.J. Klar, Materials processing using radio-frequency ion-sources: ion-beam sputter-deposition and surface treatment. Rev. Sci. Instrum. **90**, 023901 (2019)
6. M. Sugiyama, G. Sigesato, A review of focused ion beam technology and its applications in transmission electron microscopy. J. Electron Microsc. Tech. **53**, 527–536 (2004)
7. J.Y. Park, J.P. Singh, J. Lim, S. Lee, Development of XANES nanoscopy on BL7C at PLS-II. J. Synchrotron Rad. **27**, 545–550 (2020)
8. S. Reyntjens, R. Puers, A review of focused ion beam applications in microsystem technology. J. Micromech. Microeng. **11**, 287 (2001)
9. B. Mallick, Physics of ion beam synthesis of nanomaterials. in *Nanostructured Materials and their Applications* (2020), pp. 143–171
10. S. Mant, B. Holländer, D. Lenssen, M. Löken, Ion beam synthesis of silicon-based materials. Mater. Chem. Phys. **54**, 280–285 (1998)
11. K. Nagarajappa, P. Guha, A. Thirumurugan, P.V. Satyam, U.M. Bhatta, Low-energy ion beam synthesis of Ag endotaxial nanostructures in silicon. Appl. Phys. **124**, 402 (2018)
12. G. Wang, Z. Liu, S. Yang, Li Zheng, J. Li, M. Zhao, W. Zhu, A. Xu, Q. Guo, D. Chen, G. Ding, Barrier-assisted ion beam synthesis of transfer-free graphene on an arbitrary substrate. Appl. Phys. Lett. **115**, 132104 (2019)
13. A. Toma, D. Chiappe, C. Boragno, F. B.D. Mongeot, Self-organized ion-beam synthesis of nanowires with broadband plasmonic functionality. Phys. Rev. B **81**, 165436 (2010)
14. A. Bharti, R. Bhardwaj, A.K. Agrawal, N. Goyal, S. Gautam, Monochromatic X-ray induced novel synthesis of plasmonic nanostructure for photovoltaic application. Sci. Rep. **6**, 22394 (2016)
15. W. White, J. D. Budai, S. P. Withrow, J. G. Zhu, S. J. Pennycook, R. H. Magruder, D. O. Henderson, Ion beam synthesis of nanocrystals and quantum dots in optical materials. AIP Proceed. 824–827, (1996)
16. R.A. Wilhelm, E. Gruber, R. Ritter, R. Heller, S. Facsko, F. Aumayr, Charge exchange and energy loss of slow highly charged ions in 1 nm thick carbon nanomembranes. Phys. Rev. Lett. **112**, 153201 (2014)
17. S. Taller, D. Woodley, E. Getto, A.M. Monterrosa, Z. Jiao, O. Toader, F. Naab, T. Kubley, S. Dwaraknath, G.S. Was, Multiple ion beam irradiation for the study of radiation damage in materials. Nucl. Instrum. Methods B **412**, 1–10 (2017)
18. P.R. Gardner, A review of ion implantation applications to engineering materials. Mater. Des. **8**, 210–219 (1987)
19. E. Steinbauer, Fundamentals of ion-solid interactions: atomic collisions. Appl. Particle Laser Beams in Mater. Technol. **283**, 21–36 (1995)
20. J. Buchheim, R.M. Wyss, I. Shorubalko, H.G. Park, Understanding the interaction between energetic ions and freestanding graphene towards practical 2D perforation. Nanoscale **8**, 8345–8354 (2016)
21. P. Capper, S. Irvine, T. Joyce, Epitaxial crystal growth: methods and materials. in *Springer Handbook of Electronic and Photonic Materials*, ed. by S. Kasap, P. Capper (Springer Handbooks. Springer, Cham, 2017)

22. I. Petrov, V. Orlinov, S. Grudeva, On the energy efficiency of sputtering, Bulgarian. J. Phys. **18**, 214–220 (1991)
23. J.F. Ziegler, High energy ion implantation. Nucl. Instrum. Methods Phys. B **6**, 270–282 (1985)
24. D. Fink, L.T. Chadderton, Ion-solid interaction: status and perspectives. Braz. J. Phys. **35**, 735–740 (2005)
25. V.I. Shulga, A. Schinner, P. Sigmund, Effect of impact-parameter-dependent electronic energy loss on reflected-ion spectra. Nucl. Instrum. Method Phys. Section B: Beam Interactions with Mater. Atoms. **467**, 91–96 (2020)
26. A. L'hoir, S. Andriamonje, R. Anne, N.V.D.C. Faria, M. Chevallier, C. Cohen, J. Dural, M.J. Gaillard, R. Genre, M. Hage-Al, R. Kirsch, B. Farizon-Mazuy, J. Mory, J. Moulin, J.C. Poizat, J. Remillieux, D. Schmaus, M. Toulemonde, Impact parameter dependence of energy loss and target-electron-induced ionization for 27 MeV/u Xe^{35+} incident ions transmitted in [1101 Si channels. Nucl. Instruments Methods Phys. B **48**, 145–155 (1990)
27. S.T. Nakagawa, Impact parameter dependence of the electronic stopping power for channeled ions. Phys. Status Solidi B **178**, 87–98 (1993)
28. L.C. Feldmann, J.W. Mayer, S.T. Picraux, *Material Analysis by Ion Channelling* (Academic Press, New York, 1982)
29. J.F. Ziegler, J.P. Biersack, U. Littmark, *The Stopping and Range of Ions in Matter* (Pergamon Press, New York, 1985)
30. Y. Zhang, W.J. Weber, Ion irradiation and modification: the role of coupled electronic and nuclear energy dissipation and subsequent nonequilibrium processes in materials. Appl. Phys. Rev. **7**, 041307 (2020)
31. A. Gras-Marti, H.M. Urbassck, N. Arista, F. Flores, *Interaction of charged particles with solids and surfaces* (Plenum Press, New York, 1991)
32. E. Balanzat, N. Betz, S. Bouffard, Swift heavy ion modification of polymers. Nucl. Instrum. Methods Phys. Res. B **105**, 46–54 (1995)
33. M.C. Ridgway, F. Djurabekova, K. Nordlund, Ion-solid Interactions at the extremes of electronic energy loss: examples for amorphous semiconductors and embedded nanostructures. Curr. Opin. Solid State Mater. Sci. **19**, 29–38 (2015)
34. J.F. Ziegler, M.D. Ziegler, J.P. Biersack, SRIM–The stopping and range of ions in matter. Nucl. Instrum. Methods Phys. Res. B **268**, 1818–1823 (2010)

Chapter 2
Swift Heavy Ion Irradiation

2.1 Swift Heavy Ions

An ion is termed as a swift heavy ion (SHI) [1] if its mass (m) and velocity (v) satisfy the following criteria

$$m \gg m_o \quad \text{and} \quad v \gg v_o Z^{2/3},$$

where m_o, v_o, Z are the mass of electrons, the Bohr velocity and the atomic number, respectively. These ions when penetrate through the target material, losses their energy in the medium. The energy of SHI ranges from tens of MeV to a few GeV and provides a controlled variation in the physical properties of materials based on the interaction between the incident ions and the target. As these ions pass through the material, they lose their energy in two ways—one, through inelastic collision by electrons in the target which is called S_e, and second, elastic collision with target nuclei that are called S_n. At high energies, the S_e leads over S_n and it is the main reason behind the observed variations in the target material. In general, as the ions pass through the material, different types of defects are induced in the system depending upon the value of S_e and S_n. If S_e is less than a particular threshold (S_{eTh}) value, the point defects are produced and if it is larger than S_{eTh}, columnar defects are expected to occur. These SHIs are the type of particle radiations for which the S_e dominates [2, 3]. The generation of these defects is expected to bring significant modifications in the physical properties of materials (Fig. 2.1).

© The Author(s), under exclusive license to Springer Nature Switzerland AG 2022
P. Kumar et al., *Ion Beam Induced Defects and Their Effects in Oxide Materials*,
SpringerBriefs in Physics,
https://doi.org/10.1007/978-3-030-93862-8_2

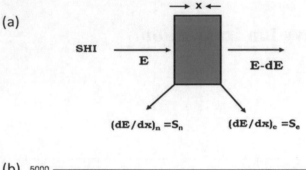

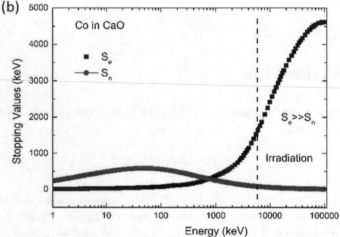

Fig. 2.1 **a** Penetration of a Swift Heavy Ions of energy (E) through the target of distance (*x*). An amount dE is lost in the target in the form of nuclear stopping, S_n and electronic stopping, S_e. **b** Variation of S_n and S_e in CaO with the energy of Co ions

2.2 Types of Accelerator

Energetic ions are generated by using particle accelerators. A particle accelerator produces a beam of charged particles that are used for diverse kinds of research. There are two basic types of particle accelerators depending upon the geometry: linear accelerators and circular accelerators. The linear accelerators propel particles linearly, while the circular accelerators propel particles around a circular track. Both these have specific advantages for the experimental work [4–6]. Some of the variants are known as electrostatic accelerator [7], cascade accelerator [8], linear resonance accelerator [9], circular electron accelerator-the betatron [10], and Van de Graff Accelerator [11]. Recent developments have been made over plasma-based charged particle accelerators [12]. In this chapter, a brief account of a typical Pelletron accelerator is presented.

2.3   The Pelletron

A typical high energy ion accelerator installed at Inter-University Accelerator Centre (IUAC), New Delhi, India is a tandem van-de-Graaff type Pelletron machine (Fig. 2.2) manufactured by Electrostatics International Inc. (EII), USA [13, 14]. This is one of the largest particle accelerators in south-east Asia.

This accelerator is capable of accelerating ions from proton to uranium up to an energy of about 200 MeV depending upon the ion. The 15 UD tandem electrostatic accelerator at IUAC is installed with a vertical configuration in an insulating steel tank which is 26.5 m long, and 5.5 m in diameter. Highly insulating toxic gas sulfur hexafluoride (SF_6) is filled at 4.0 torr pressure inside it for insulation. Inside the tank, there is a high voltage terminal and it can be charged through a high voltage that can be varied from 4 to 15 MV. The terminal is connected to the tank vertically through ceramic-titanium tubes called accelerating tubes. A potential gradient is maintained through these tubes, from high voltage to ground from the top of the tank to the terminal as well as from the terminal to the bottom of the tank. The negative ions from the ion source are injected into the accelerator and then accelerated towards the terminal. At the terminal, negative ions are stripped off a few electrons while passing through the stripper foils (e.g. carbon foils), and thereby converted to positive ions.

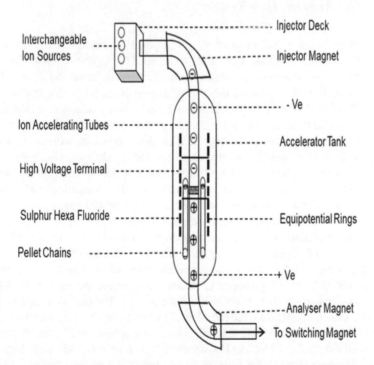

Fig. 2.2 A schematic of Pelletron accelerator installed at Inter University Accelerator Centre, New Delhi. (Adapted from https://www.iuac.res.in/pelletron-accelerator)

These positive ions are further accelerated as they proceed to the bottom of the tank at ground potential. As a result, the ions emerging out of the accelerator gain energy by $E = V_\pi (q + 1)$ MeV, where, V_π is the terminal potential (in MV) and q is the number of positive charges (charge state) on the ions after stripping. These high energy ions are then analyzed to the required energy with the help of a 90° bending magnet known as the analyzer magnet (a quarter section of 1800 mm diameter) and directed to the desired experimental area with the help of a multiport switching magnet which can deflect the beam to any one of the several beamlines in the beam hall. There are three main parts of the Pelletron Accelerator.

2.3.1   The Ion Injector System

It consists of a 380 kV injector deck with three interchangeable ion sources: a source of negative ion by Cesium sputtering (SNICS), direct extraction duoplasmatron (DED), and RF periodic charge exchange source (ALPHATROSS) [15].

2.3.2   The Accelerating System

The insulating column that supports the high potential terminal consists of thirty modules of 1 MV, 15 on either sides of the terminal. The upper portion of the column is referred to as the low energy section while the portion below the terminal is the high energy section. Two shorted sections with zero potential gradients (commonly known as the dead section) are provided one in low energy sections and the other in high energy sections, for equipment housing. Both are provided with an electron trap and a sputter ion pump. The low-energy dead section is provided with an electrostatic quadrupole lens while the high energy dead section is equipped with a second foil stripper assembly. A shorting rod system is also provided for temporarily selected column modules without entering the pressure vessel. Two insulating shafts run from each ground end to the terminal and are used to drive four 400 cps generators, which provide power for the equipment such as heaters, lenses, pumps, foil changers, etc., housed in the column dead section and terminal. The charging of a high voltage terminal to 15 MV is done by using the pelletron charging chain. There are two independent subsystems with one charging chain in each so that each chain is required to supply 100 μA. In the presence of accelerating tubes, the maximum terminal potential reached so far is 16.6 MV without beam [16]. The charging chains consist of stainless steel pellets (hence the term Pelletron) carrying charges, which are joined to one another by insulating nylon fabric connectors. There are 27 pellets/meter and the speed of the chain is 15 m/s. The terminal has an offset quadrupole triplet lens with a variable aperture and a faraday cup to select the desired charge state after stripping (Fig. 2.3).

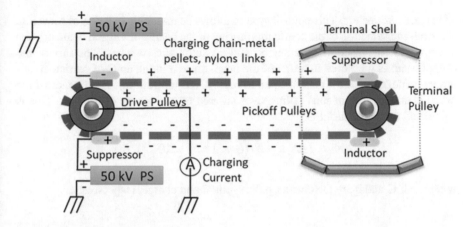

Fig. 2.3 The pelletron charging system

To produce a high-quality beam and to facilitate good experimental conditions, several other arrangements are also provided. For example, the purity of the ion beam (both in terms of mass and energy) is controlled by the injector and analyzer magnets. The injector magnet acts as a mass analyzer and sends suitable ions to the accelerator units. The analyzer magnets select the desired points and the scanner magnet helps in scanning the beam over the samples when required at a frequency of 10 Hz (x-axis) × 4 Hz (y-axis). Besides, two ion bunchers, one for light and the other for heavy ions, are also provided to facilitate beams.

2.3.3 Control System

The complete process of producing the desired ion beam demands optimization of numerous parameters. Since the ion beam is highly charged, it diverges after some distance. To ensure that most of the ions reach the experimental site, the beam is repeatedly focused and aligned at several places using electric quadrupole doublet magnet systems. These magnets are specially designed and fabricated to meet the requirement that more than 90% of ions reach the experimental site. A typical transmission probability of a proton beam from the entrance of the machine to the exit after acceleration at 15 MV accelerating potential is about 92% [13].

Another important thing is the maintenance of the vacuum. The entire tube, through which the beam travels, is maintained in a vacuum at 10^{-8}–10^{-9} torr. Automatic pneumatic valves are placed at various places so that the vacuum is not destroyed in the accidental opening of the tube. Outside the accelerating tubes, SF_6 is filled at very high pressure (~4 torrs) in the tank housing to provide insulation.

Because of all these facts, more than five hundred parameters are required for better beam delivery. All of these could not be handled manually at the same time. Hence, a computerized remote control system assists the monitoring and regulating of various parameters. This control system allows beam tuning in terms of focusing, steering, and selection of the beam; monitoring of the beam current and beam profile; vacuum system, machine parameters, and all automatic switches. The flux of the output beam is measured in terms of the beam current in the units of particle nano-amperes (pnA) (where 1 pnA is equal to 6.25×10^9 ions/cm²/s). The fluence of the beam is the number of ions falling over an area of one square centimeter. This is calculated using the formula:

$$\Phi = (\Delta I \times C)/(q \times 1.602 \times 10^{-19}).$$

where, ΔI, C and q are the current pulses, count and charge, respectively.

2.4 Material Science Beam Line

Material Science Beam line is at 15° to the right of zero degree beam-line at Inter University Accelerator Centre, New Delhi. Various experiments can be performed at this beamline apart from the temperature-dependent irradiation, the residual gas analysis (RGA), and the elastic recoil detection analysis (ERDA). Irradiation experiments are performed at high vacuum chamber evacuated to a pressure of 10^{-6} Torr. The experimental chamber used in the irradiation experiment is shown in Fig. 2.4.

To set up various parameters for irradiation, a light source along with a camera is fitted inside the chamber. Focusing of the beam is carried out by monitoring the

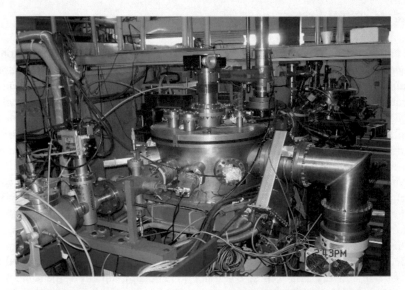

Fig. 2.4 Experimental Chamber at Materials Science Beam Line, IUAC, New Delhi, India

Fig. 2.5 Horizontal view of the sample holder used in materials science related experiments for swift heavy ion irradiation at IUAC, New Delhi, India

ionoluminescence from the quartz crystal assembled at each face of the ladder. The ion beam can be uniformly scanned over an area of 15×15 mm^2. A current integrator is used to determine the ion fluences. From the ladder, secondary electrons produced during the irradiation may contribute to this. A suppressor with a negative bias of around 120 V is used to suppress this contribution.

2.5 Sample Mounting

A typical target ladder used for the experiments related to materials science at IUAC is shown in Fig. 2.5. The main parts of the ladder are made of good quality stainless steel and oxygen-free copper. The ladder on which samples are mounted is made of copper block and it can be moved vertically remotely from the Data Room.

The copper block holding the samples could be maintained at low temperature by pouring liquid nitrogen inside the stainless steel tube attached to it. The salient features of the target ladder include the motorized vertical movement which could be controlled from the remote and the possibility of rotation through 90°, to bring the samples in front of the beam exposure. The quartz substrate at the end of the ladder is used for the focusing of the beam. This ladder can also be used for both the in-situ measurements during irradiations with the help of the electrical connection present at the edges of the ladder. The uniqueness of the ladder is that several samples can be mounted on all four sides of the rectangular-shaped ladder.

References

1. J.U. Andersen, G.C. Ball, J.A. Davis, W.G. Davies, J.S. Forster, J.S. Geiger, H. Geissel, V.A. Rayabov, Energy loss of heavy ions at high velocity. Nucl. Instrum. Methods Phys. Rev. B **90**, 104–111 (1994)

2. M. Lang, F. Djurabekova, N. Medvedev, M. Toulemonde, C. Trautmann, Fundamental phenomena and applications of swift heavy ion irradiations. Comprehens. Nuclear Mater. **1**, 485–516 (2020)
3. L. Douillard, J.P. Duraud, Swift heavy ion amorphization of quartz-a comparative study of the particle amorphization mechanism of quartz. Nucl. Instrum. Methods Phys. Res. B **107**, 212–217 (1996)
4. G.P. Averyanov, A.V. Kobylyatskiy, The charged particle accelerators subsystems modelling. J. Phys.: Conf. Ser. **798**, 012154 (2017)
5. O.B. Malyshev, Vacuum in particle accelerators: modelling, design, and operation of beam vacuum systems, vacuum in particle accelerators (Vol. 523, 2020)
6. Y.M. Ado, High-energy charged-particle accelerators. Sov. Phys. Usp. **28**, 54 (1985)
7. V.N. Glazanov, Electrostatic charged-particle accelerators. The Soviet J. Atomic Energy **6**, 98–106 (1961)
8. A. Gulevich, V. Chekunov, O. Fokin, O. Komlev, O. Kukharchuk, C. Melnikov, N. Novikova, L. Ponomarev, E. Zemskov, Concept of electron accelerator-driven system based on subcritical cascade reactor. Prog. Nucl. Energy **50**, 347–352 (2008)
9. H. Wiedemann, Circular accelerators, particle accelerator physics, pp. 59–80. (2015)
10. C.J. Karzmark, N.C. Pering, Electron linear accelerators for radiation therapy: history, principles and contemporary developments. Phys. Med. Biol. **18**, 321 (1973)
11. R.J. Van de Graaff, A 1,5000,000 volt electrostatic generator. Phys. Rev. **38**, 1919 (1931)
12. R. Bingham, J.T. Mendonça, P.K. Shukla, Plasma-based charged-particle accelerators, plasma phys. Control. Fusion **46**, R1 (2003)
13. G.K. Mehta, A.P. Patro, 15UD pelletron of the nuclear science center-status report. Nucl. Instr. Methods Phys. B A268 334–336 (1988)
14. N. Madhvan, Nuclear science centre-an advanced accelerator-based Inter-University research center at New Delhi. Resonance **2**, 92–96 (1997)
15. D. Kanjilal, S. Chopra, M.M. Narayanan, S. Iyer, V. Jha, R. Joshi, S. K. Dutta, Testing and operation of the 15 UD pelletron at NSC. Nucl. Instr. Methods A **328**, 97–100 (1993)
16. D. Kanjilal, in *Proceedings Symposium of North Eastern Accelerator Personnel* (Kansas State University, pp. 33, 1990)

Chapter 3
Low Energy Ion Implantation

Several important processes are involved in the implantation of ions into solids which in turn alter the physical states of the near-surface region. A clear understanding of these processes and the modification in the structure is invaluable for utilizing ion implantation phenomena to engineer surfaces through the controlled modification of surface properties. In the basics of ion implantation processes, fluence, stopping powers, ion ranges, damage energy, enhanced diffusion, ion mixing, and sputtering are discussed. The collisional and solid-state processes combine to establish the composition profile as well as the physical state of the implanted material. A qualitative understanding of these processes helps to clarify some of the general considerations which are required when one decides to alter the near-surface properties of any material via ion implantation.

The fundamental understanding of ion implantation was developed during the 1960s. Since that time, this technique has been widely utilized both as a research tool as well as a practical method to control certain surface properties of solids. The ion implantation technique provides a very effective depth and spatial selection of implanted species that cannot be achieved by any other chemical and physical techniques. This is because selected energetic ions (with the energy of a few keV) go up to a certain depth and are implanted into the target matrix that modifies the target material properties due to their presence. The variations in the ion fluence and energy provide freedom to implant desired concentration/amount of selected ions into the host material. When the ions enter into the solid, they continuously lose energy and changes direction by collisions with the target atoms (see Fig. 3.1). At low incident ion energies (few keV), it is recognized that the atomic cascades arising out of the primary, secondary, and higher-order knocked-on atoms mean that a 'cascade' produces voids, extended defects, or disordered/amorphous zones in all three dimensions [1–3]. The collision processes are then frequently referred to as 'ballistic'. Figure 3.1a shows the schematic for the formation of a collision cascade in the implantation process while Fig. 3.1b displays the trajectory profile (calculated by TRIM software) of 80 keV Ni ions into the CeO_2 matrix [4]. It is clear from

© The Author(s), under exclusive license to Springer Nature Switzerland AG 2022   17
P. Kumar et al., *Ion Beam Induced Defects and Their Effects in Oxide Materials*,
SpringerBriefs in Physics,
https://doi.org/10.1007/978-3-030-93862-8_3

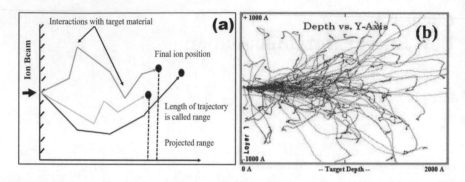

Fig. 3.1 **a** Schematic of ion implantation in the target **b** Profile of 80 keV Ni ions into CeO_2 [4]

Fig. 3.1 that impinging ions collide with target atoms that in turn deviate from their path and form a collision cascade.

3.1   Ion Implantation

The ion implantation technique has emerged out to be the best-suited technique in the semiconductor industries, as it provides precise control over the depth and spatial selection of species that cannot be achieved by any other chemical / physical technique. For this process, a focused beam of ions having only a single atom or molecule is required. It should also exhibit the well-defined value of energy. For implantation, preference is given to the ionized particles, as they can be easily accelerated. When the energy transfer from the incoming ions to the host lattice atom is high enough, a so-called primary knock-on atom is kicked from its lattice site. In turn, this primary knock-on atom can displace other nearby atoms (secondary knock-on atoms) and so on, thus creating an atomic collision cascade. After a series of inelastic electronic collisions and a few elastic knock-on collisions during the high energy regime of the ion trajectory, the momentum of the projectile drops drastically. As the ion traverses in the low energy regime, the nuclear stopping power that has an inverse square relationship with the ion energy, becomes the dominant stopping mechanism. This is because at low energies, the projectile spends more time in the vicinity of the target nuclei, and hence the cross-section of ballistic nuclear collisions becomes greater. The relative contributions of the energy loss mechanisms depend on the energy of the incoming ions and can be estimated with the help of the SRIM software.

When the ion passes through the lattice, it displaces the host atoms from their regular lattice site, resulting in the creation of defects. These defects can only be tempted when the energy transfer from the incoming ions exceeds the threshold energy. The number of defects and the relative lattice disorder (i.e. the degree of crystallinity) after ion implantation depend on several parameters such as mass and energy of the implanted ions, ion current density, implantation angle, total implanted fluence, substrate temperature, and of course on the radiation hardness of the material. This section summarizes those processes which determine the depth distributions of the atomic composition and the primary defect generation in the implanted solid.

3.1.1 Implantation Fluence

The implantation fluence is the number of ions implanted per unit area and is often referred to as the term "dose". For uniform scanning of the beam, it can be accurately determined by measuring the charge that each ion carries to the target. However, when an ion hits a surface then several secondary electrons, photons, and sputtered atoms are emitted, which can strike other surfaces and release additional secondary electrons. Thus, for beam current measurements, secondary electron suppression is often used but this procedure is not as accurate as having the sample surrounded by a Faraday cup with only a small solid angle available for the beam to enter [5, 6]. Besides, secondary electrons from beam-defining slits need to be excluded from the Faraday cup (e.g. using biased shields or plates). Finally, in electrostatically swept systems requiring high uniformity, there is often a static deflection in the beam direction at the last sweep plates and the vacuum is kept at moderately low values ($\sim 10^{-7}$ torr) to minimize beam non-uniformities due to charge exchange neutrals. The absolute beam fluence measurements of accuracy $\sim 1\%$ are achievable with all the necessary precautions [5, 6].

In addition to accuracies associated with the experimental system, some ions will be reflected out of the target through a sequence of atomic collisions before they come to the rest position. It is noted that ion reflection becomes significant when the incident ion is lighter than the target atoms. This reflection depends primarily upon the mass ratio between the incident and target atoms, the incident angle, and the dimensionless energy parameter (ε) [7].

$$\varepsilon = \frac{0.8853a_0}{Z_1 Z_2 e^2 \left(Z_1^{2/3} + Z_2^{2/3}\right)^{1/2}} \frac{M_2}{M_1 + M_2} E. \tag{3.1}$$

Here, a_0 is Bohr's radius, Z is the atomic number, M represents mass, E is the energy, and subscripts 1 and 2 are referred to as projectile and target atoms, respectively. The reflection is also found to increase with the energy reduction. This reflection will not be detected in a Faraday cup arrangement, but it has been well described experimentally and can be predicted theoretically [8]. Both reflection and sputtering increase substantially where the beam is incident at glancing angles to the surface. This would become an important consideration in terms of the loss of implanted species, for example, in rotating a cylindrical part in front of the beam. The loss of the implanted species due to sputtering and the migration of the implanted ions will be considered in subsequent sections.

3.1.2   Stopping Power

As stated in Chap. 1, the stopping power can be partitioned into a nuclear component (due to elastic collisions with the nuclei) and an electronic component (due to inelastic energy loss to the electrons) [9, 14].

Nuclear stopping usually dominates the energy loss rate at lower energies whereas electronic stopping dominates at higher energies. In typical regions of interest for ion implantation, both these contributions are substantial. In addition to determining the ion ranges, nuclear stopping is important for the description of sputtering and displacement damage production. Electronic stopping is important in electronic excitation, which results in secondary electron emissions and can produce defects in the insulators. It is based on integrating the elastic energy losses in the collisions [9]. The electronic stopping is proportional to the ion velocity for the low energy regime (ion velocities $\ll v_0 Z_2^{2/3}$, the Thomas Fermi velocity of the target electrons) [9, 10]. In this regime, the electronic stopping can be thought of as that of a slowly moving ion in an electron gas, for which the force on the ion is proportional to its velocity. The electronic stopping for any ion target combination [9] is then given approximately by $S_e = kE^{1/2}$, where.

$$
k = \frac{0.0793 Z_1^{\frac{2}{3}} Z_2^{\frac{1}{2}} \left(M_1 + M_2^{3/2} \right)}{\left(Z_1^{\frac{2}{3}} + Z_2^{\frac{2}{3}} \right)^{3/4} M_1^{3/2} M_2^{1/2}}.
\tag{3.2}
$$

The value of k is typically ~ 0.15 for implantation conditions. In addition to the general dependence of the electronic stopping on Z_1 and Z_2 as given in Eq. 3.2, there is an oscillatory contribution with the atomic number of the projectile Z_1' which correlates with electron shell effects and may be thought in terms of dependence on the ion size [11, 12]. These oscillations are of the order of 30% in an amorphous target or the absence of channelling. Channelling usually occurs when an energetic ion is steered by crystal atom rows or planes through a series of gentle correlated collisions. For channelled particle trajectories, the stopping power and the resulting ion ranges can be strongly altered from the values described above for the non-channelling case. For implanted atoms to be channelled, they must typically be incident angles within $1°$ to $10°$ of a low index direction. During channelling, close encounter interactions with the nuclei $\ll 0.1$ Å are prevented and the projectile spends most of its time in a region of low electron density. Thus, the total stopping power can be greatly reduced. Besides, the Z_1 oscillations in the stopping power can be quite large [13].

3.1.3   Ion Range Distributions

The total path length of an ion can be estimated by taking the total electronic and nuclear contributions to the stopping power and integrating the reciprocal of the loss rate per unit depth from the incident energy to the zero energy. The actual projected range relative to the incident direction of the ion is smaller due to the lateral deflections caused by the collisions as the ion slows to rest. The ratio of the projected to the total ion range depends on the mass ratio of the projectile and the target atoms. The transport equations were derived by Lindhard et al. [9], which finally provide the distribution of ions. In the first order, the ion range distribution is well described by the Gaussian and the values of the first two moments for various ion-target combinations as a function of energy have been tabulated by several researchers [14–18]. Higher moments in the range distribution have also been considered and various schemes are developed for obtaining distributions based on these moments [17]; their effect on the distribution is often small. The second moment is referred to as the range straggling and values can be obtained for straggling along the incident direction of the ion beam as well as in the transverse direction.

3.2   Accelerator for Ion Implantation

A typical ion implantation system consists of an ion source, an ion acceleration column, a mass separation system, a region for shaping and sweeping the beam, and finally an implantation chamber for holding and manipulating the target under vacuum. The schematic diagram of a typically low-energy ion beam facility housed at Inter University Accelerator Centre (IUAC), New Delhi, India is shown in Fig. 3.2.

In the low-energy ion beam facility at IUAC, New Delhi, there are three different beam lines, out of these one beam line is dedicated to the Material Science facility. The ion source forms a plasma that contains an appreciable fraction of ions of the atomic species to be implanted. The ions are extracted from the source and accelerated through an electrostatic potential gradient with appropriate ion optics. Mass separation is typically accomplished with a magnetic field to select a particular mass ion species. Beam sweeping is often done electrostatically to provide uniform area implantation. Usually, the beam current to the target is integrated to determine the number of implanted atoms per unit area. Typical implantation energies range from 10 to 100's of keV, with typical penetration depths ranging from 1 to 100's of the nanometer (nm). There are film deposition processes, such as ion milling, which also utilize energetic ions. In this case, only a fraction of the incident atoms or atom clusters are ionized and accelerated with the average energy per atom several orders of magnitude lower than in the case of ion implantation. At present, there appears to be an increasing trend towards combining ion implantation and film deposition processes. Such combined processes may well perceive substantial future applications, particularly in the metals where thicker layers of substantial alloy composition and strong substrate adherence are desirable.

The picture of the Low Energy Ion Beam Facility (LEIBF) at Inter-University Accelerator Centre (IUAC), New Delhi, India is shown in Fig. 3.3. This facility consists of an Electron Cyclotron Resonance (ECR) ion source (Nanogun from Panteknik) installed on a high voltage deck. All the electronic control devices of

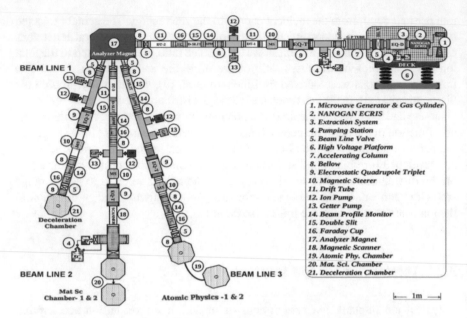

Fig. 3.2 Schematic diagram of low energy ion beam facility at Inter University Accelerator Centre, New Delhi, India. [Adapted from www.iuac.res.in]

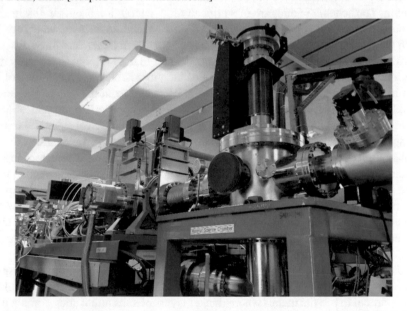

Fig. 3.3 Low energy ion beam facility at Inter University Accelerator Centre, New Delhi, India

the ECR source including the high-power UHF transmitter (10 GHz), are placed on a high voltage deck. These are controlled through optical fiber communication in multiplexed mode. This facility provides multiple charged ion beams at a wide range of energies (a few keV to about ~1 MeV) for experiments in various disciplines: Atomic Physics, Molecular Physics, and Materials Sciences. Apart from this, there is another accelerator known as the Negative Ion Implanter beam Facility (NIIBF) based on the source of negative ions by Cesium sputtering (SNICS) at IUAC dedicated to delivering metal beams of 30 to 200 keV and current from few nA to μA.

References

1. K.E. Sickafus, E.A. Kotomin, B.P. Uberuaga, in *Radiation Effects in Solids* (Springer, 2004)
2. E. Knystautas, in *Engineering thin films, and nanostructures with ion beams*, (CRC Press an imprint of Taylor & Francis Group, 2005)
3. S. Dhara, Formation, dynamics and characterization of nanostructures by ion beam irradiation. Crit. Rev. Solid State Mater. Sci. **32**, 1 (2007)
4. SHI irradiation induced modification in CeO_2 thin films and Nanostructures, M. Varshney Ph.D. thesis, submitted to Dr. B. R. Ambedkar Univeristy, Agra (2012)
5. J. L'Ecuyer, J.A. Davies, H. Matsunami, How accurate are absolute Rutherford backscattering yields. Nucl. Instr. Meth. **160**, 337 (1979)
6. S. Matteson, M.A. Nicolet, Electron and ion currents relevant to accurate current integration in McV ion backscattering spectrometry. Nucl. Instr. Meth. **160**, 301 (1979)
7. R. Kossowsky, S.C. Singhal, Surface engineering surface modification of materials. in *Proceedings of the NATO Advanced Study Institute on Surface Engineering*, Les Arcs, France, July 3–15 (1983)
8. J. Bottiger, J.A. Davies, P. Sigmund, K.B. Winterbon, On the reflection coefficient of keV heavy-ion beams from solid targets. Radiat. Eff. **11**, 133 (1971)
9. J. Lindhard, M. Scharff, H.E. Schiott, Mat. Fys. Medd., Dan. Vide Selsk. **33**, 14 (1963)
10. O.B. Firsov, Soviet Phys. JETP **9**, 1076 (1959)
11. P. Hvelplund, B. Fastrup, Stopping cross section in carbon of 0.2–1.5-MeV atoms with $21 \leq Z1 \leq 39$. Phys. Rev. **165**, 408 (1968)
12. K.B. Winterbon, Z1 oscillations in stopping of atomic particles. Can. J. Phys. **46**, 2429 (1968)
13. F.H. Eisen, G.J. Clark, J. Bottiger, J.M. Poate, Stopping power of energetic helium ions transmitted through thin silicon crystals in channeling and random directions. Radiat. Eff. **13**, 93 (1972)
14. J.A. Davies, L.M. Howe, in *Site Characterization and Aggregation of Implanted Atoms in Materials*, NATO Advanced Study Series B, ed. by A. Perez, R.Coussemen (1980)
15. J.F. Gibbons, W. Johnson, S.W. Mylroie, in *Projected Range Statistics* (1975)
16. D.K. Brice, in *Ion Implantation Range and Energy Deposition Distributions* (1975)
17. K.B. Winterbon, in *Ion Implantation Range and Energy Deposition Distributions* (1975)
18. U. Littmark, J.F. Ziegler, Range distributions for energetic ions in all elements (1980)

Chapter 4
Consequences of Heavy Ions and Models

In the previous chapters, it was mentioned that the interaction of energetic ions with the materials lead to several processes in the system depending upon the energy, ion fluence, and the physical properties of the materials themselves. These processes behave differently depending upon the energy regimes: (i.e. high energy regime (in MeV range) and low energy regime (in keV range)). This chapter describes briefly various effects in these energy regimes, models, and the defects created/induced based on these effects.

4.1 High Energy Regime

In the high energy regime, the induced effects are explained based on two models: The Coulomb Explosion model and the Thermal Spike model [1].

4.1.1 The Coulomb Explosion Model

When fast heavy ions pass through the target material, neighboring positive target ions are produced, which are mutually repulsive [2]. The amount of ionization can be estimated based on a purely classical consideration of the ionization cross-section by taking into account the various ionization energies of the target atoms and the Coulomb interaction between the screened projectile and the target electrons [3]. If the time of the projectile to cover a distance of one atomic diameter a_o is short in comparison to the response time of the conduction electrons, the wake of the projectile is a long cylinder of radius a_o containing charged target ions. The schematic diagram of the Coulomb repulsion model is represented in Fig. 4.1. In the left figure, the arrow symbols (green colour symbols) shows the ion path while the right figure

© The Author(s), under exclusive license to Springer Nature Switzerland AG 2022
P. Kumar et al., *Ion Beam Induced Defects and Their Effects in Oxide Materials*,
SpringerBriefs in Physics,
https://doi.org/10.1007/978-3-030-93862-8_4

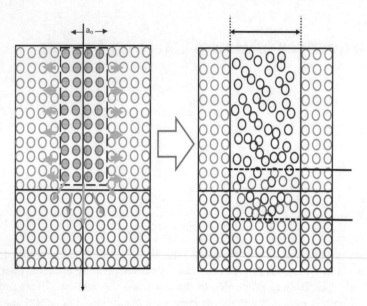

Fig. 4.1 Schematic diagram showing the Coulomb explosion model. The arrows in the left figure show the ion path while the right figure displays the atomic displacement region as well as the mixed zone region after irradiation [4]

shows the atomic displacement region. The right-side figure also displays the mixed zone region after irradiation [4]. This cylinder explodes radially under the Coulomb force until the ions within the cylinder are screened by the conduction electrons. After screening, the matter surrounding the projectile trajectory is mechanically polarized as a result of the movements of the ions and relaxes "gradually". The occurrence of ion beam-induced deformation (IBID) is due to the difference in the duration of the Coulomb explosion and the resultant mechanical polarization. Since these two time periods are comparable in cylindrical metal, plastic deformation is not observed in these systems [5, 6].

4.1.2 Thermal Spike Model

Most of the theoretical work concerning the formation of the latent track results from the thermal spike model making it the most accredited damage model. A thermal spike is a high-temperature region formed along the trajectory of an energetic ion. If electronic energy loss (S_e) is sufficiently high then the material (crystalline or amorphous) is melted within a cylinder of radius R, and it cools down within 10^{-12}–10^{-11} s with a cooling rate of 10^{15}–10^{14} K/s. This process may result in the formation of an amorphous phase [7, 8] completely different from the original material. According to the thermal spike model, the system is considered as the combination of the electronic subsystem and the lattice subsystem [9, 10]. The incoming ions transfer their energy to the electrons in 10^{-17} s and the electrons reach an equilibrium state in a time of the order of 10^{-15} s. This energy is transferred to the atomic lattice by electron–phonon coupling. Therefore, in this thermodynamic system, having a volume V, pressure P, and temperature T, excess heat is generated during heavy ion bombardment. The local temperature of the system increases by an amount of ΔT and the heat is transferred to the lattice in the form of energy via the electron–phonon coupling. Kaoumi et al. [11] proposed a model to describe grain growth under irradiation in the temperature-independent regime, based on the direct impact of the thermal spikes on grain boundaries. In the model, grain-boundary migration occurs by atomic jumps within the thermal spikes, biased by the local grain-boundary curvature driving. The jumps in the spike are calculated based on Vineyard's analysis of thermal spikes and activated processes using a spherical geometry for the spike [10]. This model incorporates cascade structure features such as subcascade formation and the probability of these subcascades occurring at the grain boundaries.

4.2 Low Energy Regime

4.2.1 Damage Energy Deposition and Ion Cascades

The partitioning of the energy loss rate of an ion into electronic and nuclear components is also vital in determining the damage distributions due to implantation. The energy loss into nuclear collisions can result in the atomic displacements inside the crystal for energy transfers greater than the threshold energy, which is typically ~25 eV (for isolated atomic displacements). Thus, as the ion slows down to rest, atom displacement results in the production of vacancy-interstitial pairs and more complex defects. The depth distribution of the energy into damage can be obtained theoretically [12, 13].

The equations for the moments of distribution have been derived from the transport equations in the same way as for the ion distributions. These equations take into account knock-on atoms in the target as they further redistribute their recoil energy between electronic and nuclear processes [14]. The damage distributions generally differ substantially from a Gaussian profile and methods have been developed to reconstruct the distribution from the low order moments [15]. A direct method has also been developed which solves the ion depth distributions at all intermediate energies and determines the transfer of energy and accounts for the recoil contribution at each point [16].

4.2.2   Radiation Enhanced Diffusion and Segregation

The migration of implanted atoms and the redistribution of solute atoms may occur after the collision cascade formation due to point defect motion [17–19]. For example, in crystals where atomic diffusion occurs via vacancy motion, the local concentration of vacancies will be enhanced by the ion damage cascade. Then at that temperature sufficient for vacancy motion during or after implantation, enhanced diffusion of implanted solutes can result. Besides, the defect flow can also lead to the radiation-enhanced segregation of solute atoms. This occurs when there is a coupling of positive energy between the solute and the defects so that the net flow of defects from the generation region (ion cascade) to annihilation sites (surface, precipitates, grain boundaries) results in a preferential solute flow. Radiation enhanced diffusion broadens concentration gradients and simply depends on an enhanced concentration of mobile vacancies and interstitials. In contrast, radiation enhanced segregation locally concentrates on solutes and depends on a specific direction of net defect flow. Both effects are based on the free motion of vacancies and/or interstitials induced by the ion cascade and can be readily observed under appropriate implantation conditions.

4.3   Ion Beam Induced Effects in Materials

For the present discussion, we distinguish these relatively well-defined mechanisms of radiation enhanced diffusion and segregation from other ion cascade effects leading to atomic redistribution; the latter is discussed in the next section under the heading of ion beam mixing.

4.3.1   Surface Modification

Modification of surfaces is one of the important aspects of ion beam tools [20–22]. The change of surface roughness for Pt [23] and SiO_2 [24] by low energy ion beam is shown in Fig. 4.2. A similar effect is also observed for MgO thin film under Zn and Fe ion implantation [25]. Based on these results and the literature survey, it is concluded that ion beam implantation plays an important role in the modification of surface roughness.

Similar behavior of surface modification is observed for MgO thin films under 120 MeV Ag ions [26] and Au thin films [27] under [58]Ni and [107]Ag ions (Fig. 4.3). Yasui et al. investigated the surface modification of MgO thin films under oxygen irradiation [28].

Thus, both ion implantation and irradiation affect the surface roughness of the thin films, however, a direct correlation between the surface roughness and fluence of the

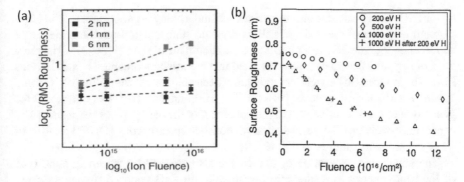

Fig. 4.2 **a** Root mean square (RMS) roughness in nm with ion fluence for Pt thin films of thicknesses 2, 4, and 6 nm [23] and **b** Surface roughness of SiO_2 films are shown as a function of fluence under different energies of hydrogen ions [24]

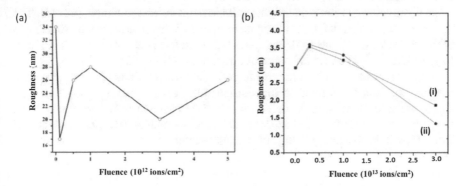

Fig. 4.3 **a** Surface roughness of MgO thin films under 120 MeV Ag ions [26] (CC-BY 4.0) and **b** Variation of the roughness of Au thin films under (i) 58 MeV Ni and (ii) 107 MeV Ag ions [27]

ion beam could not be established. In the case of bilayers [29, 30] and multilayers [31], the modifications of interfaces are also being observed. The ion beam can also promote mixing at the interface [32]. This may lead to the evolution of new phases at the interfaces [33]. It is also pertinent to mention here that ion implantation can also lead to the formation of bilayers in the single thin film [34]. The various examples exhibiting all these phenomena are discussed in the next chapter.

4.3.2 Defect Formation

Defect creation is the general process of ion beam techniques [35]. Though information on these defects can be revealed from the simulation using the TRIM (Transport of Ions in Matter) calculations, experimental studies play an important role in identifying the nature of these defects. Experimental studies using photoluminescence show $(Zr_{Al}-V_N)^0$ defects in Zr implanted aluminum nitride (AlN) [36, 37]. The presence of oxygen vacancies is also revealed in ZnO implanted using O^+ and B^+ ions [38]. The n-type ZnO single crystals have been implanted with 500 keV O^+ and 1.2 MeV Zn^+ ions using doses between 1×10^{11} and 2×10^{12} ions cm^{-2}, and the generation of deep-level defects in the upper part of the bandgap has been studied by capacitance–voltage (C-V) and deep level transient spectroscopy (DLTS) performed up to sample temperatures of 500 K [39].

The nature and generation of defects are also affected by the energy and type of ion beam energy. A systematic investigation in ZnO nanorod implanted using H^+ ions shows the generation, evolution, and annihilation of defects with a dose of implantation (Fig. 4.4) [40]. The point-defect relaxation increases with ion fluence in MgO using irradiation with 1.2 MeV Au^+ ions [41]. Thus, these tools are effective in engineering defects in different materials.

Generally, oxygen ion irradiation results in point defects in oxides [42–44]. The presence of columnar defects is observed in zinc ferrite nanoparticles when irradiated with 200 MeV Ag^{15+} ions [45]. A similar type of defect is also observed in $LaTaO_3$ [46] and Al_2O_3 [47] under influence of different ions (Fig. 4.5).

Thus, ion beam tools are effective to produce different kinds of defects in oxide material that can be utilized to tune various physical properties of these oxides. The systematic variation in the defects that leads to the creation/enhancement in the magnetic properties of various oxide semiconductors such as TiO_2, ZnO, MgO, and $ZnFe_2O_4$ will be discussed in the subsequent chapter.

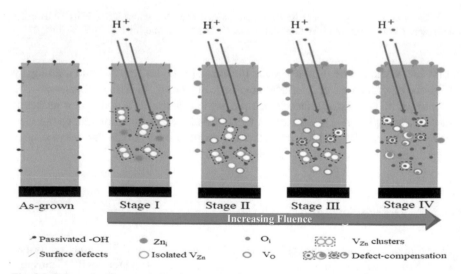

Fig. 4.4 Schematic diagrams indicating the generation, evolution, and annihilation of defects in ZnO nanorod arrays under 180 keV H^+ ion beam irradiation [40]

(a) 13.5 keV/nm@2.15MeV/u
Ni^{19+} irradiation; E$_{ele}$-Peak region

(b)

Fig. 4.5 Latent track observation in LaTaO$_3$. panels **a** and **b** are used to show these tracks at different magnification [46]

References

1. E.M. Bringa, R.E. Johnson, Coulomb explosion, and thermal spikes. Phys. Rev. Lett. **88**, 165501 (2001)
2. A.I. Ryazanov, S.A. Pavlov, E.V. Metelkin, A.V. Zhemerev, Effect of Coulomb explosion on track formation in metals irradiated by heavy ions. J. Exp. Theor. **101**, 120–127 (2005)
3. S. Klaumiinzer, M.D. Hou, G. Schumacher, Coulomb explosions in a metallic glass due to the passage of fast heavy ions?. Phys. Rev. Lett. 850–853 (1986)
4. I.P. Jain, G. Agarwal, Ion beam induced surface and interface engineering. Surf. Sci. Rep. **66**, 77 (2011)
5. A. Audouard, R. Mamy, M. Toulemonde, G. Szenes, L. Thomé, Impacts of GeV heavy ions in amorphous metallic alloys investigated by near-field scanning microscopy. EPL **40**, 527 (1997)
6. S. Klaumünzer, Plastic deformation of amorphous solids by track overlap. Int. J. Radiation Appl. Instrument. Part D. Nuclear Tracks Radiation Measurem. **19**, 91–96 (1991)
7. G. Szenes, Ion-velocity dependent track formation in yttrium iron garent: a thermal spike analysis. Phys. Rev. B **52**, 6154–6157 (1995)
8. G. Szenes, Thermal spike model of amorphous track formation in insulators irradiated by swift heavy ions. Nucl. Instrum. Methods Phys. Res. B **116**, 141–144 (1996)
9. F. Seitz, On the disordering of solids by the action of fast massive particles. Discuss, Faraday Soc. **5**, 271 (1949)
10. S. Ghosh, A. Gupta, P. Ayyub, N. Kumar, S.A. Khan, D. Banerjee, R. Bhattacharya, Swift heavy ion irradiation-induced damage creation in nanocrystalline Li–Mg ferrite thin films. Nucl. Instrum. Methods Phys. Res. B **225**, 310–317 (2004)
11. D. Kaoumi, A.T. Motta, R.C. Birtcher, A thermal spike model of grain growth under irradiation. Int. J. Appl. Phys. **104**, 073525 (2008)
12. D.K. Brice, Ion implantation depth distributions: energy deposition into atomic processes and ion locations. Appl. Phys. Lett. **16**, 103 (1970)
13. T. Endo, A. Sadaki, S. Sasaki, K. Sawa, T. Wada, Spatial distributions of damages introduced into gap by collimated MeV-electron beam irradiations (II): Comparison with electron distributions by scattering. Radiat. Effect. **84**(2011), 89–105 (1984)
14. W.B. Winterbon, P. Sigmund, J.B. Sanders, Mat. Fys. Medd., Dan. Vide Selsk. **37**, 14 (1970)
15. K.B. Winterbon, *Ion Implantation Range and Energy Deposition Distributions, Vor:-2* (IFI/Plenum, NY, 1975)
16. D.K. Brice, in Ion Implantation Range and Energy Deposition Distributions. vol 1. (IFI/Plenum, N~1975)

17. S.M. Myers, Ion-beam-induced migration and its effect on concentration profiles. Nucl. Instr. Meth. **168**, 265–274 (1980)
18. A.D. Marwick, Physics of ion implantation (Ion Cascade Processes and Physical State of the Implanted Solid). Nucl. Instr. Meth. f82/183, 827 (1981)
19. L.E. Rehn, in *Metastable Materials Formation by Ion Implantation*, ed. by S.T. Picraux, W.J. Choyke (North Holland, NY, 1982) pp. 17
20. F. Frost, R. Fechner, B. Ziberi, J. Völlner, D. Flamm, A. Schindler, Large area smoothing of surfaces by ion bombardment: fundamentals and applications. J. Phys.: Condens. Matter. **21**, 224026 (2009)
21. M. Kumar, R.K. Pandey, P. Rajput, S.A. Khan, U.B. Singh, D.K. Avasthi, A.C. Pandey, SHI induced surface re-organization of non-amorphisable nanodimensional fluoride thin films. Phys. Chem. Chem. Phys. **19**, 23229–23238 (2017)
22. K. Bhattacharjee, S. Bera, D.K. Goswami, B.N. Dev, Nanoscale self-affine surface smoothing by ion bombardment and the morphology of nanostructures grown on ion-bombarded surfaces. Nucl. Instrum. Methods Phys. B **230**, 524–532 (2005)
23. M. Kumar, R.K. Pandey, S. Pathak, Vandana, S. Ojha, T. Kumar, R. Kumar, Surface engineering of Pt thin films by low energy heavy ion irradiation. Appl. Surf. Sci. **540**, 148338 (2021)
24. E. Chason, T.M. Mayer, Low energy ion bombardment induced roughening and smoothing of SiO$_2$ surfaces. Appl. Phys. Lett. **62**, 363 (1993)
25. J.P. Singh, W.C. Lim, J. Lee, J. Song, K.H. Chae, Surface and local electronic structure modification of MgO film using Zn and Fe ion implantation. Appl. Surf. Sci. **432**, 131–139 (2018)
26. J.P. Singh, I. Sulania, J. Prakash, S. Gautam, K.H. Chae, D. Kanjilal, K. Asokan, Study of surface morphology and grain size of irradiated MgO thin films. Adv. Mater. Lett. **3**, 112–117 (2012)
27. P. Dash, P. Mallick, H. Rath, A. Tripathi, J. Prakash, D.K. Avasthi, S. Mazumder, S. Varma, P.V. Satyam, N.C. Mishra, Surface roughness and power spectral density study of SHI irradiated ultra-thin gold films. Appl. Surf. Sci. **256**, 558–561 (2009)
28. N. Yasui, H. Nomura, A. Ide-Ektessabi, Characteristics of ion beam modified magnesium oxide films. Thin Solid Films **447–448**, 377–382 (2004). https://doi.org/10.1016/s0040-6090(03)01087-3
29. A. Gupta, D. Kumar, Interface modification in Fe/Cr epitaxial multilayers using swift heavy ion irradiation. Nucl. Instrum. Methods Phys. Res. B **244**, 202–205 (2006)
30. E.G. Njoroge, C.C. Theron, J.B. Malherbe, N.G. van-der-Berg, T.T. Hlatshwayo, V.A. Skuratov, Surface and interface modification of Zr/SiC interface by swift heavy ion irradiation. Nucl. Instrum. Methods Phys. Res. B. **354**, 249–254 (2015)
31. J.P. Singh, W.C. Lim, K.H. Chae, S. Gautam, K. Asokan, Swift heavy ion irradiation-induced effects in Fe/MgO/Fe/Co multilayer. Mater. Des. **101**, 72–79 (2016)
32. S. Kraft, B. Schattat, W. Bolse, S. Klaumünzer, F. Harbsmeier, A. Kulinska, A. Löffl, Ion beam mixing of ZnO/SiO$_2$ZnO/SiO$_2$ and Sb/Ni/Si interfaces under swift heavy ion irradiation. J. Appl. Phys. **91**, 1129 (2002)
33. S. Gupta, D.C. Agarwal, J. Prakash, S.K. Tripathi, S. Neeleshwar, B.K. Panigrahi, D.K. Avasthi, Synthesis of PbTe thermoelectric film by high energy heavy ion beam mixing. AIP Conf. Proc. **1393**, 343–344 (2011)
34. R. Bhardwaj, B. Kaur, J.P. Singh, M. Kumar, H.H. Lee, P. Kumar, R.C. Meena, K. Asokan, K.H. Chae, N. Goyal, S. Gautam, Role of low energy transition metal ions in interface formation in ZnO thin films and their effect on magnetic properties for spintronic applications. Appl. Surf. Sci. **479**, 1021–1028 (2019)
35. K. Kanemoto, F. Imaizumi, T. Hamada, Y. Tamai, A. Nakada, Dependence of ion implantation: induced defects on substrate doping. Int. J. Appl. Phys. **89**, 3156 (2001)
36. A. Aghdaei, R. Pandiyan, B. Ilahi, M. Chicoine, M.E. Gowini, F. Schiettekatte, L.G. Fréchette, D. Morris, Engineering visible light-emitting point defects in Zr-implanted polycrystalline AlN films. Int. J. Appl. Phys. **128**, 245701 (2020)

37. J.B. Varley, A. Janotti, C.G. Van-de-Walle, Defects in AlN as candidates for solid-state qubits. Phys. Rev. B. **93**, 161201 (2016)
38. Z.Q. Chen, Annealing process of ion-implantation-induced defects in ZnO: chemical effect of the ion species. Int. J. Appl. Phys. **99**, 093507 (2006)
39. L. Vines, J. Wong-Leung, C. Jagadish, E.V. Monakhov, B.G. Svensson, Ion implantation induced defects in ZnO. Physica B Condens. **407**, 1481–1484 (2012)
40. T. Wu, A. Wang, L. Zheng, G. Wang, Q. Tu, B. Lv, Z. Liu, Z. Wu, Y. Wang, Evolution of native defects in ZnO nanorods irradiated with a hydrogen ion. Sci. Reports **9**, 17393 (2019)
41. D. Bachiller-Perea, A. Debelle, L. Thomé, J.P. Crocombette, Study of the initial stages of defect generation in ion-irradiated MgO at elevated temperatures using high-resolution X-ray diffraction. J. Mater. Sci. **51**, 1456–1462 (2016)
42. P. Pandey, Y. Bitla, M. Zschornak, M. Wang, C. Xu, J. Grenzer, D.C. Meyer, Y.Y. Chin, H.J. Lin, C.T. Chen, S. Gemming, M. Helm, Y.H. Chu, S. Zhou, Enhancing the magnetic moment of ferrimagnetic $NiCo_2O_4$ via ion irradiation driven oxygen vacancies. APL Mater. **6**, 066109 (2018)
43. A.B. Mei, S. Saremi, L. Miao, M. Barone, Y. Tang, C. Zeledon, J. Schubert, D.C. Ralph, L.W. Martin, D.G. Schlom, Ferroelectric properties of ion-irradiated bismuth ferrite layers grown via molecular-beam epitaxy. APL Mater. **7**, 111101 (2019)
44. K.N. Rathod, K. Gadani, D. Dhruv, V.G. Shrimali, S. Solanki, A.D. Joshi, J.P. Singh, K.H. Chae, K. Asokan, P.S. Solanki, N.A. Shah, Effect of oxygen vacancy gradient on ion-irradiated Ca-doped $YMnO_3$ thin films. J. Vac. Sci. Technol. **38**, 062208 (2020)
45. J.P. Singh, G. Dixit, H. Kumar, R.C. Srivastava, H.M. Agrawal, R. Kumar, Formation of latent tracks and their effects on the magnetic properties of nanosized zinc ferrite. J. Magn. Magn. Mater. **352**, 36–44 (2014)
46. X. Han, Y. Liu, M.L. Crespillo, E. Zarkadoula, Q. Huang, X. Wang, P. Liu, Latent tracks in ion-irradiated $LiTaO_3$ crystals: damage morphology characterization and thermal spike analysis. Curr. Comput.-Aided Drug Des. **10**, 877 (2020)
47. J.H. O'Connell, R.A. Rymzhanov, V.A. Skuratov, A.E. Volkov, N.S. Kirilkin, Latent tracks and associated strain in Al_2O_3 irradiated with swift heavy ions. Nucl. Instrum. Methods Phys. Res. B **374**, 97–101 (2016)

Chapter 5
Defect Driven Magnetic Properties of Oxide Materials

In this chapter, the effect of defects via ion beam techniques on the properties of some well-known oxide semiconductors viz. titanium oxide (TiO_2), zinc oxide (ZnO), magnesium oxide (MgO), and zinc ferrite ($ZnFe_2O_4$) are discussed. To date, many efforts have been devoted in search of high Curie temperature based diluted magnetic semiconductors (DMS) such as transition metal ion (Ti, V, Cr, Mn, Fe, Co,) doped ZnO, TiO_2, SnO_2, In_2O_3, MgO, and HfO_2 materials applicable in spintronic devices [1–7]. The study starts with the observation of magnetic behavior in these oxides via doping with transition metal ions. But, these dopants create controversies among the scientific community as the origin of magnetism is unclear. Some researchers have argued that magnetism is due to the doping of magnetic ions while others claimed that magnetism is associated with the creation of defects in the host matrix [8–10]. Some research groups later argued that magnetism might be associated with cluster formation among the dopant ions. Therefore, various efforts have been put to clarify this strange behavior in oxides. This chapter focuses on understanding the magnetic behavior in these oxide semiconductors induced by defects produced by ion beams.

5.1 Magnetism in Titanium Dioxide (TiO_2)

In the past few decades, research interest in TiO_2 has grown exponentially as evident from a growing number of publications in these materials. TiO_2 is a very interesting and versatile material with a wide range of applications, including use in microelectronics due to its high dielectric constant and in optical coatings due to its high refractive index [11–16]. It also has excellent optical transmittance in visible and near-infrared regions. TiO_2 is an important material for the nanoscale application and it has unique optical and electrical properties [17–22]. It is an interesting material for UV absorption, due to the scientific and technological applications in dye-sensitized solar cells, photocatalysis, spintronics and sensing [19–22].

© The Author(s), under exclusive license to Springer Nature Switzerland AG 2022
P. Kumar et al., *Ion Beam Induced Defects and Their Effects in Oxide Materials*,
SpringerBriefs in Physics,
https://doi.org/10.1007/978-3-030-93862-8_5

Understanding the origin of ferromagnetism at room temperature (RT) in the undoped wide bandgap semiconductors is one of the most important challenges in the condensed matter physics and physics of magnetism. Since the observation of RT ferromagnetism in these systems, [23–31] considerable attention has been paid to the phenomenon so-called d^0 magnetism [13]. It has been suggested by different groups that magnetism in these systems is induced either due to anion/cation vacancies, di-Frenkel pairs, and or quantum-confinement effects [27–49]. Esquinazi et al. used the density functional theory (DFT) calculations to estimate the magnetization and X-ray magnetic circular dichroism (XMCD) values [30]. However, the experiments on the observed magnetism in the oxide materials produced quite controversial results and are still debated in the literature because of the origin of the magnetism in these systems, which are observed by different kinds of defects [29].

Swift heavy ion (SHI) irradiation is an important way to modify the properties of the material by introducing structural disorder, defects, and/or columnar amorphization depending upon the extent of the electronic energy loss mechanism in the system [32]. Besides, it has been shown that SHI can also induce crystalline-to-crystalline phase transition instead of amorphization and damage creation [33].

Rath et al. [34] have observed a transformation of phase from anatase to rutile after irradiating with 200 MeV Ag ions in the TiO_2 thin films. In the X-ray diffraction (XRD) pattern of films, both (anatase and rutile) phase is observed in the pristine sample. The effect in the intensity of (101) peak is also not observed up to a fluence of 1×10^{12} ion/cm^2. At the next fluence (3×10^{12} ions/cm^2) of irradiation, however, the anatase peak is suppressed considerably, while the rutile phase on the other hand shows a complex variation with irradiation fluence. The intensity of (110) peak is observed to decrease at the initial dose (5×10^{11} ions/cm^2) of irradiation where the additional peak corresponding to tetragonal δ-TiO_2 is observed at 38.1° while the peaks in hexagonal phases of TiO_2 appear at 44.3° and 45.1°. There seems to be a distribution of the integrated peak intensity between the rutile phase and the two unstable phases of titanium oxide at this fluence of irradiation. The intensity of XRD peaks recorded decreases considerably at 1×10^{12} ion/cm^2 due to the unstable phases. At the highest fluence (3×10^{12} ions/cm^2), the peak due to the rutile phase is most intense and peaks due to all other phases are suppressed. Further, the rutile phase seems to be grain oriented in the unirradiated films as well as in films irradiated at the highest dose, while at intermediate fluences, a rutile peak corresponding to (211) reflection appears. In the Raman spectra of TiO_2 thin films, the pristine films have peaks related to both the anatase and rutile phases as well as related to the Si substrate peaks. No remarkable change occurs in the anatase phase upon irradiating the films up to the fluence of 1×10^{12} ions/cm^2 as shown in the XRD results. At the fluence of 3×10^{12} ions/cm^2, the E_g, B_{1g}, and B_{1g}, A_{1g} Raman modes of anatase phase are suppressed, while the intensity of the peaks due to B_{1g}, E_g, and A_{1g} modes of rutile structure is increased [35]. Raman spectra also confirmed the XRD result of the transformation from anatase to rutile phase at this fluence of irradiation. This change in the phase transformation can be explained by the applicability of the thermal-spike model, where 200 MeV Ag ions lead to a local temperature rise along their path and nucleate rutile phase well inside the grains of anatase phase instead of at their grain boundaries. The SHI technique has a unique post-deposition treatment for the formation of pure rutile TiO$_2$ phase due to the intense interaction of the incident ion with the target atoms,

Thakur et al. reported the role of SHI on the properties of TiO$_2$ thin films prepared by radio frequency magnetron sputtering on sapphire substrates [36] by 200 MeV Ag^{+15} ions at different fluences. The calculated value of the lattice parameters of the pristine film was found to be higher compared to the bulk value. It is indicating a strain in the lattice of these films. An additional diffraction peak at 42.3° corresponding to the reflection from the (410) plane of the brookite phase of TiO$_2$ is also observed at irradiation fluence 1×10^{11} ions/cm^2. This characteristic peak has shown growth in the intensity with irradiation fluence of 1×10^{12} to 5×10^{12} ions/cm^2 along with the progression of another reflection from the (121) plane of the brookite phase of the TiO$_2$ which is dominating at the highest fluence. The expansion in the TiO$_2$ lattice with irradiation is observed. It is also noted that the reflections from the anatase phase are suppressed while a diffraction peak from the (111) plane of the rutile phase appears at the highest fluence (5×10^{12} ions/cm^2). These results suggest that SHI induces a controlled structural disorder and/or phase transition from anatase to brookite-rutile phase in the TiO$_2$. This is in confirmation of the fact that instead of amorphization and damage creation, SHI can also create crystalline-to-crystalline phase transitions.

The dominating structure of the brookite phase in the SHI irradiated film could have an important implication on its electronic and magnetic properties. The pristine film is shown to have paramagnetic behavior at RT, while the irradiated film demonstrates the saturation behavior confirming the ferromagnetic nature of the TiO$_2$ system [37]. Therefore, a structural phase transition from anatase to the mixture phase of brookite and rutile of TiO$_2$ with increasing irradiation fluence followed by a significant distortion in the TiO$_6$ octahedra. X-ray absorption spectroscopy (XAS) experiments carried out at O K and Ti L$_{3,2}$ edges display significant changes in the electronic structure due to modifications in the unoccupied Ti 3d levels by hybridization of O 2p-Ti 3d.

The structural and magnetic properties of TiO$_2$ thin films prepared by electron beam evaporation technique, annealed at 900 °C, and irradiated with 500 keV Ar^{2+} ions were investigated at different fluences of 1×10^{14} to 5×10^{16} ions/cm^2 [38]. The XRD patterns of different films are presented in Fig. 5.1a. The most intense peak observed at 25.35° in the film either corresponds to (101) of anatase phase TiO$_2$ or (120) of brookite phase TiO$_2$. However, it is observed that the intensity of the peak at 25.35° increases in the film with increasing the irradiation fluence from 1×10^{14} to 5×10^{16} ions/cm^2. The diffuse diffraction peaks in film irradiated by 1×10^{14} to 5×10^{16} ions/cm^2 indicate amorphization. The appearance of the Ti$_4$O$_7$ phase observed here would be due to the increasing annealing temperature.

Raman measurement is done for confirmation of the phase, which is shown in Fig. 5.1b. Pristine and 1×10^{14} ions/cm^2 irradiated films show the Raman peaks at 144, 396, and 637 cm^{-1}, which correspond to E$_g$, B$_{1g}$, and E$_g$ modes of anatase phase. The film irradiated at 5×10^{16} ions/cm^2 shows peaks at 151 and 634 cm^{-1} that

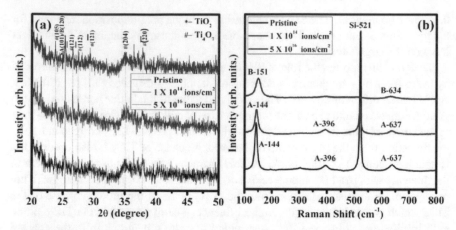

Fig. 5.1 **a** Effect of ion fluences on the XRD results of TiO$_2$ thin films irradiated by 500 keV Ar^{2+} ions **b** Raman spectra of films at different ion fluences [38]

correspond to the A$_{1g}$ mode of the brookite phase. XRD and Raman results show the transformation from anatase to brookite phase at the fluence of 5×10^{16} ions/cm^2.

A peak in the Raman spectrum is observed at 521 cm^{-1} related to Si substrate. The Si-related peak has been suppressed with ion fluence. It is indicating the creation of defects in silicon substrate after irradiation. Thus, the low-energy ion irradiation not only induces a phase transformation from anatase to brookite in TiO$_2$ thin film but also damages the silicon substrate beneath the film because the projected range of the ion is higher than the thickness of the film. Among three existing phases of TiO$_2$ such as rutile, anatase, and brookite, it may be mentioned that while anatase and rutile are the most common phases of TiO$_2$, the brookite phase is rarely observed in TiO$_2$. Bharati et al. reported a transformation from anatase to brookite phase of TiO$_2$ under low energy ions i.e. 500 keV Ar^{2+} ions, where the nuclear energy loss (S$_n$) induced processes play a dominant role [38].

Magnetic properties have been studied in TiO$_2$ thin films by using the field (H) dependent magnetization (M) measurement as depicted in Fig. 5.2 (Inset shows the zoomed view of hysteresis curves) [38]. The variation in the magnetic field is observed from 0 to 10 kOe with the variation of the fluencies of irradiation. All the films show hysteresis behavior with finite coercivity and remanence which is a signature of ferromagnetism. The magnetization saturates at about 6 kOe with saturation magnetization 11.37, 7.26, and 6.27 emu/cc for pristine, 1×10^{14} ions/cm^2, and 5×10^{16} ions/cm^2, respectively. For pristine and 1×10^{14} ions/cm^2 irradiated film, the coercivity is 61 Oe, while remanence is found to be 1.08 and 0.45 emu/cc, respectively. For 5×10^{16} ions/cm^2 irradiated film, the coercivity is 34 Oe and remanence is 0.32 emu/cc. The occurrence of room-temperature ferromagnetism (RTFM) in all the films is due to the presence of the Ti$_4$O$_7$ phase where Ti is in a +3 oxidation state. As Ti^{3+} is magnetic, ferromagnetism would be due to the presence of Ti$_4$O$_7$ in the structure, which is confirmed by the XRD results. The

Fig. 5.2 Magnetization as a function of applied magnetic field for pristine and irradiated TiO$_2$ film. The inset shows the zoomed view of the M-H loops [38]

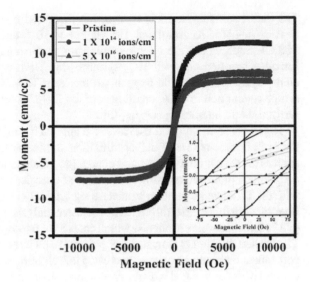

magnetic moment in the pristine film is higher than the irradiated samples due to higher oxygen vacancies. It provides the information that irradiation decreases the concentration of defects.

Mohanty et al. reported the structural and magnetic properties of Co-doped TiO$_2$ thin films irradiated with 100 MeV Ag^{7+} ions [39]. The presence of a loop at RT confirms the ferromagnetic behavior in the TiO$_2$ films irradiated at different fluences. It is evident that although coercivity remains almost the same, the saturation magnetization (M$_s$) decreases with an increase in ion fluence. Thakur et al. reported the ferromagnetism by Ag irradiation of energy 200 MeV in undoped paramagnetic TiO$_2$ thin film followed by a phase transformation from anatase to rutile and brookite mixed-phase [37]. The ferromagnetism in irradiated samples is explained based on distortion in TiO$_6$ octahedra due to ion irradiation.

RTFM has been also observed in a rutile TiO$_2$ polycrystalline sample after 4 MeV Ar^{5+} ion irradiation reported by Sanyal et al. [40]. The implantation depth is calculated by the computer code SRIM software [41] and found to be 2.48 μm. To obtain a vacancy concentration, one has to specify a displacement threshold energy in SRIM by displacement threshold energy (E$_d$) values of Ti and O in TiO$_2$; these values have been taken as 57 eV [42]. It is seen that S$_e$ is much lower than the columnar defect production threshold (tens of keV/nm) in TiO$_2$ [43]. Therefore, only vacancies and vacancy clusters are expected in the target material at the earlier stage of research on the magnetism of TiO$_2$ but later include other defects like Ti-Frenkel pairs are also included in the study. It is observed that Ar ions are most effective at displacing oxygen atoms.

The M versus H curve is measured at room temperature for the Ar irradiated TiO$_2$. The hysteresis loop for the Ar irradiated portion of the sample is prominent when the diamagnetic background coming from the unirradiated portion of the rutile

TiO_2 has been subtracted [44] from the measured curve. The saturation magnetization field for Ar irradiated TiO_2 is $\sim 4 \times 10^{-4}$ emu/g with the coercive field 168 Oe. Such values are very low compared to ferromagnetic ZnO but are typical for other reported undoped TiO_2 systems [45]. High-energy Ar ions are very efficient to incorporate stable oxygen vacancies in TiO_2. Oxygen vacancies promote n-type conduction in TiO_2, creating partially filled 3d orbital in Ti atoms (Ti^{3+}), and ferromagnetic interaction sets in [45].

Robinson et al. reported the extensive approach adopted enables examination of the probabilistic nature of defect formation and the precise extraction of quantities such as the threshold displacement energy [46]. Calculations of the elemental defect proportions indicated a distinct difference at energies around displacement energy, with the Ti defect population in anatase significantly larger than the oxygen defects. Investigating this further through defect cluster analysis highlighted the accumulation of a Ti defect complex in anatase which consists of two vacancies and two interstitials. The labeled di-Frenkel pair, stability is found to be induced by the localization of the constituent Frenkel pairs, which is created in isolation spontaneously recombine. The energy barrier associated with the recombination of the di-Frenkel pair suggests annihilation. Further, the reason behind the ferromagnetism in TiO_2 films was reported by many research groups [47–51]. Rumaiz et al. [47] attributed the ferromagnetism in TiO_2 to oxygen vacancies rather than Ti^{3+}/Ti^{2+} cations. However, Wei et al. [48] explain that 2p electrons of oxygen play an important role in the exchange interaction and ferromagnetic ordering. The presence of an oxygen vacancy is associated with two electrons which may localize the neighboring Ti ions transforming into Ti^{3+} or may be delocalized in the TiO_2 matrix. Kim et al. [49] have observed RTFM in both anatase and rutile phases of TiO_2. The higher magnetic moment has been attributed to more oxygen defects in the distorted TiO_6 octahedra. Higher magnetization in the pristine film could be due to higher oxygen vacancies, which was obtained from the X-ray photoelectron spectroscopy technique (XPS).

Botsch et al. reported a detailed investigation about the ferromagnetism in TiO_2 [31]. Most experimental investigations of artificial FM phases emerging upon ion irradiation reported in the literature were performed at high ion energies > 100 keV. Figure 5.3 shows the evolution of the remanent magnetic moment, m_{rem}, measured in samples S200 (irradiated with Ar^+ ions at $E_{ion} = 200$ eV) and S1000 (irradiated with Ar^+ ions at $E_{ion} = 200$ eV) as a function of the irradiation fluence f. By fitting the data to a power law [$m_{rem} \propto (f - f_0)^\beta$], we find critical fluences $f_0 = 5 \times 10^{15}$ ions cm^{-2} and 2.5×10^{16} ions cm^{-2} for samples S200 and S1000, respectively. The resulting critical exponents are $\beta = 0.22 \pm 0.04$ and 0.42 ± 0.07, respectively. Comparing these exponents to the theoretical values, they observed that the remanence observed in sample S1000 follows the critical behavior of a bulk three-dimensional (3D) percolation transition, while sample S200 follows the critical behavior of a magnetic bilayer system. These results match very well with the aforementioned thickness of the FM phases, that predicted the emergence of a magnetic bilayer in sample S200, while in sample S1000 the FM phase spans over eight layers. So, The magnetism is increased with increasing the fluences for both samples. It is indicated that SHI may be used as a tool for tailoring the structural, electronic, and even magnetic properties of the

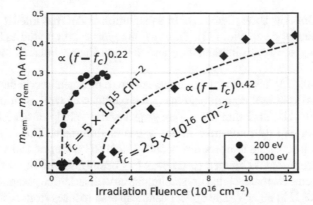

Fig. 5.3 Remanent magnetic moment m_{rem} at zero field, measured at T = 300 K after setting a magnetic field B = 5 T, as a function of the irradiation fluence f. The background remanence $m_{0\,rem}$ of the unirradiated samples was subtracted. The symbols represent experimental data of samples S200 (●) and S1000 (■). The dashed lines represent fits to $m_{rem} \propto (f - fc)^\beta$, with the critical fluences (fc) and critical exponents (β) as indicated [31]

TiO₂ system. These magnetic properties are attributed to the oxygen vacancies as well as di-Frenkel pairs in the TiO₂ structure, created by ion beam irradiation, which is responsible for magnetism in TiO₂.

5.2 Magnetism in Zinc Oxide (ZnO)

Zinc oxide (ZnO) is another important material that gained significant importance after the observation of magnetic behavior in various oxide semiconductors. Among oxides, ZnO has gained a lot of research interest owing to their applications such as varistors [49], lasers [50, 51], field-effect transistors [52], high sensitive chemical sensors [53, 54], piezoelectric transducers. The high Curie temperature (above room temperature) of ZnO makes it suitable for spintronic applications [55–58]. Besides, ZnO is an *n*-type semiconductor having a wide direct bandgap of 3.37 eV with relatively 60 meV free exciton binding energy, which makes it suitable for commercial applications [59–61].

The ferromagnetism in pure ZnO is closely related to defects or structural disorders. It was proposed that defects like Zn and O vacancies [62–66], Zn interstitial [67, 68], grain boundaries [69, 70], and lattice distortion [71] can induce ferromagnetism in pure ZnO. Oxygen vacancies (V_O) as defects not only modify the valence of transition metal ions but modulate the band structure of the host matrix. Based on the existing literature, it has been found that these defects can easily be introduced by ion beam techniques. SHI and low-energy ion beam techniques are proven to be effective approaches to tailor the defect concentration in the host matrix. However, there is still some debate on the origin of RTFM in ZnO, whether magnetism is induced by intrinsic, secondary phases, defects, or by substitution as ZnO has filled d-orbitals that rule out its magnetic behavior [72]. This section reports various studies highlighting the low-energy ion beam implantation and SHI irradiation-induced magnetism in ZnO.

Kumar et al. [73] have shown that implantation of nitrogen ions in the host matrix plays a crucial role in the modifications in the bandgap as well as in the magnetic properties. For this, ZnO thin films were grown with the help of the RF sputtering technique over Si (100) substrate in an Ar gas environment. The deposition was carried out at a substrate temperature of 500 °C and RF power of 150 W for half an hour. The highly c-axis oriented (002) peak has been observed in all the samples. XRD measurements indicate that crystallite size as well as lattice parameter rise with ion fluences of N at 60 keV. The optical band gap also reduces from 3.27 to 3.04 eV with the incorporation of nitrogen ions having an ion fluence of 1×10^{17} ions/cm^2. It is proposed that the mixing of shallow N 2p states with the valence band of zinc oxide causes a reduction in the bandgap. Furthermore, the saturation magnetization (measured at room temperature) approximately doubles at this concentration of nitrogen ions in the host lattice (Fig. 5.4). The decrease in bandgap can be correlated to the improvement in magnetic character due to the induced hybridization by N ions over the O site. The experimental results of enhancement in saturation magnetization and reduction in bandgap with the incorporation of N ions are also seen in the density functional theory (DFT). The angle of implantation was kept fixed at 90° for different ion fluence [73].

Hariwal et al. [74] reported that ferromagnetism in these systems can also be tailored by varying the implantation angle. ZnO thin films were deposited by RF sputtering technique with RF power of 200 W for 30 min without heating the Si substrate. This results in the growth of 300 nm thick films as evident from the AFM measurements. The ion beam implantation is carried out by altering the thin film orientation (30°, 60° and 90°) concerning the incident ion beams. XRD measurements imply that 60° implantation angle samples (ZnO: N60) (002) peak is most intense

Fig. 5.4 Magnetization curves for Pristine ZnO and 5×10^{16} & 1×10^{17} ions/cm^2 N implanted ZnO thin film [73]

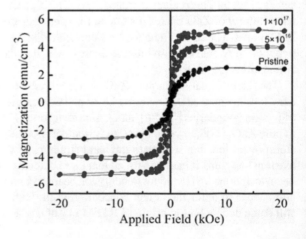

among all other samples. They refer to the improvement in the crystallinity with the formation of grains and grains boundaries in the matrix with the addition of N ions. The crystallite size & dislocation density are maximum while the surface roughness and strain are the minima for 60° implantation angle sample. The bandgap value decreases for this implantation angle. This has been attributed to the higher order of mechanical stress near dislocations due to the formation of N_2O bands which acts as a source of O as well as N. The incorporation of N ions results in the formation of Zn-N bonds due to which there is an enhancement in 2p states and reduction in the oxygen vacancies in the system that brings the band edges close to each other resulting in the reduction in the bandgap. The saturation magnetization for the ZnO: N60 sample is maximum and it is about six times that of pure ZnO thin film.

Kumar et al. [75] proposed another effective way to tailor the optical bandgap and magnetic properties concurrently with the implantation of carbon ions (C ions with 100 keV energy) in ZnO thin films. Here, thin films by deposited with RF sputtering technique on Si substrate in Ar gas at RF power of 120 W and substrate temperature of 400 °C. A highly c-axis oriented (002) peak has been observed for the complete set of samples. It is noteworthy that the position of (002) peaks varies upon the C ion fluence. It shifts to the lower 2θ side for initial fluence (5×10^{16} ions/cm^2) and the higher 2θ side for an ion fluence of 1×10^{17} ions/cm^2. Based on these observations, it is found that C ions occupy a substitutional position for the dilute concentration of ions in the host lattice, while a large concentration of these ions causes them to occupy interstitial sites. The saturation magnetization of $\sim$6.2 emu/cm^3 is observed at 5×10^{16} ions/cm^2, which is almost 3.6 times higher than pristine ZnO film, while $\sim$4.6 emu/cm^3 saturation magnetization is found at 1×10^{17} ions/cm^2. Based on their results, they demonstrate that a lower bandgap induces stronger magnetization while a reserve trend is obtained in the situation of a wider bandgap.

Another report by Bhardwaj et al. [76] elucidates the role of low energy transition metal ions such as Co, Ni, and Cu for the interface formation, which in turn is related to the induction of magnetic properties for spintronics applications. The selected beams have an energy of 100 keV at five diverse fluences (from 1×10^{15} to 5×10^{16} ions/cm^2) so that implanted ions will be able to penetrate up to the depth of $\approx$ 44 nm. It has been discussed in the introduction chapter of this book that an ion beam can be used to create an interface to tailor the material properties. In that study, researchers have shown one more application of the ion-beam implantation technique. The mechanism of bilayer formation and their interface properties has been investigated with the help of versatile techniques. XRD measurements show a systematic 2θ shift in the most intense (002) peak of ZnO revealing the substitution of the implanted ion at Zn-site. With the help of the Near-edge x-ray absorption fine structure (NEXAFS) spectroscopy, the charge state, as well as the local electronic environment, can be identified. This measurement infers that, O K-edge NEXAFS measurements for Ni-ZnO/ZnO/Si bilayer are highly sensitive to incident beam angles while no spectral change is noticed at the Zn L-edge measurements. The magnetic measurements showed that the films are ferromagnetic at room temperature and their origin of ferromagnetism can be understood through defect mediated

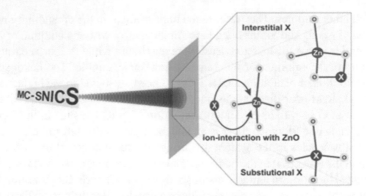

Fig. 5.5 Proposed mechanism for negative ion beam interaction X (X = Co, Ni, and Cu) with ZnO, where X can undertake one of two mechanisms of either substituting the Zn-site or going in the interstitial position of ZnO [76]. Here SNICS is referred to Source of Negative Ions by Cesium Sputtering.

bound magnetic polaron model. The schematic representation of the occupancy of implanted ions in the host matrix is shown in Fig. 5.5.

Lorite et al. [77] investigated the persistence of room temperature ferromagnetism in H^+ implanted (300 eV energy) Li doped ZnO microwires (1, 3 & 7%) synthesized by the carbothermal method. The saturation magnetization can be seen in 3 and 7% Li doped ZnO microwires only after H implantation. It is important to mention here that the M_S upsurges with H^+ implantation time and the maximum value are obtained after 90 min. This confirms that it is possible to enhance the saturation magnetization to a certain concentration of external dopant. The *Ab-initio* calculations indicate that the magnetic moment values for the Li-V_{Zn} after H^+ implantation are found in between 1 and ~2 μB. This evidence was further confirmed by XMCD measurements performed for 1% (non-magnetic) and 3% (magnetic) samples. There is no signal in the X-ray Magnetic Circular Dichroism (XMCD) measurement for non-magnetic samples while an indistinct signal appears in the range 533–540 eV for magnetic samples. It was concluded that implantation of H^+ ion at low energies in the ZnO matrix provides the solution of two different experimental problems. The first one is that the magnetic order arising from a V_{Zn} has a magnetic moment ~2 μB and their concentration is equal to the order of Li content. Furthermore, implantation of H^+ ion creates necessary defect complexes to stabilize the concentration of zinc vacancies produced.

The role of Ni ion implantation on the magnetic properties of ZnO thin films was investigated by Srivastava et al. [78]. ZnO thin films (800 nm thick) were grown over sapphire substrates using pulsed laser deposition technique and then implanted with 200 keV Ni^{2+} ions with 6×10^{15} ions/cm^2 (2% Ni conc.) and 2×10^{16} ions/cm^2 (7% Ni conc.). XRD measurements infer the growth along (002) direction and indicate the absence of any other peak corresponding to crystalline Ni or NiO phases. It is observed that the saturation magnetization is higher at lower doses of Ni ions (2%) compared to a higher one (7%). There is no direct indication of an XMCD signal for both the films within the noise level. However, if it still exists, it is not associated with the XMCD signal of ferromagnetic Ni metal or clusters. Based on X-ray linear Dichroism (XLD) measurements and Ni ion concentration, it is evident that Ni substitution in the host ZnO matrix is somewhat larger (approx. 1.3 times) for 7% sample compared to 2% sample. The leftover amount of Ni might reside in the host lattice in the form of metal clusters or its oxide in the amorphous (NiO) form. The irregular distribution of Ni ions and the possibility of formation of the NiO phase reduces the net magnetic moment of the host lattice. After a year, they analyzed the effect of annealing on the magnetic properties and corelated it with the local electronic structure [79]. XAS measurements affirm the absence of Zn cluster but provide an indication of clustering of Ni metal in the implanted films. However, the absence/reduction of this phase is noticed in the air annealed film. The formation of a larger number of bound magnetic polarons and the disappearance of the secondary phase deliver higher saturation magnetization.

The above studies are mainly focused on the role of low-energy ion beam implantation. In the subsequent step, a few examples of the importance of SHI in the ZnO matrix are presented. Defects controlled ferromagnetism is investigated in 500 keV Xenon ion irradiated zinc oxide thin films [80]. The as-deposited ZnO thin films exhibit ferromagnetic character, which further enhances with the irradiation of Xe ions up to the fluence of 2×10^{17} ions/cm^2. This character reduces with a further increase in ion fluence. Satyarthi et al. have also shown that Xenon ion beam is not able to induce the magnetic behaviour in pristine ZnO single crystals as well as in the irradiated single crystals up to an ion fluence of 3.5×10^{17} ions/cm^2. X-ray photoelectron spectroscopy measurements infer that oxygen vacancy related defects proportions are much higher in as-deposited and irradiated polycrystalline ZnO films as compared to the ZnO single crystals. Therefore, this work inferred that oxygen vacancies are the dominating factor that governs the room temperature ferromagnetic character in the samples in the present case. However, the role of Zn vacancies cannot be ignored as they can trigger magnetism even if their concentration is low.

Kumar et al. [81] highlighted the study on SHI irradiated Mg-doped ZnO thin films grown using RF sputtering technique. The defects in the deposited films were induced by 200 MeV Ag^{15+} ion beam irradiation. The structural measurements reveal the growth of highly c-axis oriented thin films. The crystallite size initially decreases with the ion fluence of irradiation whereas enhancement is observed for the highest fluence. Magnetic measurements performed at room temperature (Fig. 5.6) show that pristine $Zn_{0.9}Mg_{0.1}O$ thin-film displays ferromagnetic order, which deteriorate initially and then upsurges at the highest fluence. Therefore, depending upon the fluence variation, one could tune the defect concentrations in the host lattice, which ultimately results in the variation in magnetic properties of the samples.

Jayalakshmi et al. [82] focused their study on SHI-induced modifications with Ag ions of 50 MeV in structural, optical, and magnetic properties of pure ZnO, 5% doped vanadium, and 10% doped vanadium doped ZnO films grown over the sapphire substrate by RF magnetron sputtering. It is reported that no phase transformation and absence of secondary phase occurs in films after SHI irradiation. XRD analysis discloses that there is a minor change in the peak intensity and the

Fig. 5.6 Magnetization
curves of Pristine, 5×10^{12},
and 1×10^{13} ions/cm^2
irradiated $Zn_{0.9}Mg_{0.1}O$ thin
films. The inset shows the
magnified view of the
hysteresis curve of the
sample irradiated with an ion
fluence of 5×10^{12} ions/cm^2
[81]

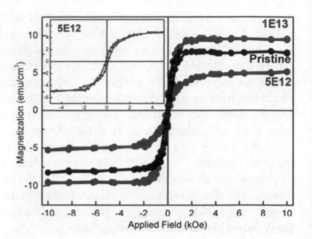

FWHM upon irradiation. The crystallite size changes from 21 to 9 nm for pure
and 5% V doped ZnO and upon irradiation crystallite size was found to be 19 and
8 nm respectively. The reduction in the size is associated with the transfer of large
amounts of energy within a short time interval that causes rapid quenching of the
material and crystallization might take place within the matrix. PL measurements
show that defect-related emission increases upon irradiation which is attributed to
the enhancement of oxygen vacancies. With the passage of high energetic ions, electronic energy stopping deteriorates oxygen bonds in the ZnO, giving rise to oxygen
vacancies (V_O). Furthermore, all the samples display the presence of a ferromagnetic character in these films. It has been argued that the rise in oxygen vacancies
upon ion beam irradiation combined with V ion concentration causes more bound
magnetic polarons (BMPs) and leads to the enhancement in ferromagnetic behavior
in irradiated V doped ZnO films.

Kumar et al. [83] studied the effect of Ag^{15+} ion beam having 200 MeV energy
on Co implanted ZnO thin films of thickness ~400 nm deposited over the sapphire
substrate using plasma-assisted molecular beam epitaxy (PAMBE) technique. In the
first step, 80 keV Co ions with varying fluences at 300 °C were implanted in the host
matrix to recover the implantation damage. XRD measurements in these films show
the presence of Co clusters due to the peak at ~44.3°. The intensity of these peaks
increases with implantation dose indicating the growth of Co cluster size. These
implanted samples were further irradiated with a 200 MeV Ag^{15+} ion beam. It has
been reported that the Co clusters can be dissolved with this beam having 1×10^{12}
ions/cm^2 fluence. The magnetization of Co implanted at 5×10^{16} ions/cm^2 and the
SHI irradiated ZnO thin film exhibits ferromagnetism at room temperature. At low
concentration of Co ions, i.e., at 2.5×10^{20} Co ions/cm^3 and 5×10^{20} Co ions/cm^3,
the saturation magnetization values are 1.28 μ_B/Co and 1.5 μ_B/Co, respectively. On
increasing Co doping, saturation magnetization decreases and reach 0.45 μ_B/Co at
1.25×10^{21} Co/cm^3. It was observed that ferromagnetism at room temperature is not
associated with the cluster formations but occurs due to the substitution of Co ions

over the Zn site. This can be understood based on energy transfer in target material through SHI irradiation. Similarly, Güner et al. [84] studied the origin of RTFM in Co implanted ZnO, and also many other authors reported on the investigation of magnetism in Fe, Mn, C, Cr, and Ni-doped ZnO [85–90].

Another report by Kumar et al. has focused upon a 200 MeV Ag^{15+} ions beam on Co-implanted ZnO thin films [91]. ZnO thin films were deposited on crystalline sapphire substrate by plasma-assisted molecular beam epitaxy technique having a thickness of 400 nm. In the first step, these ZnO thin films were implanted with 80 keV Co ions at a fluence of 1×10^{16} ions/cm^2, and then implanted films were irradiated with 200 MeV Ag^{15+} ions having fluence of 1×10^{12} ions/cm^2. XRD measurements confirm the formation of Co clusters after implantation. However, XRD measurements on irradiation infer that Co clusters are dissolved into the ZnO matrix after SHI irradiation. The semiconducting behavior is shown by the films while studying conduction mechanisms in these films. Further, the influence of SHI irradiation can be seen that the resistivity of implanted films reduces by half corresponding to ~38 mΩ at room temperature. The coercivity of ~ 65 Oe is observed in each case, indicating ferromagnetic nature at room temperature. Though, it was found that the saturation magnetization is reduced with SHI irradiation and no significant change in coercivity. The maximum magnetization for the pristine sample is recorded to be 2.80 emu/cm^3, whereas it is decreased to 2.21 emu/cm^3 for the irradiated sample. This reduction in magnetization is due to defects induced by SHI irradiation. The increase in ferromagnetism can be ascribed to the presence of oxygen vacancies generated by the SHI irradiation as well as the formation of oxide phases of Co in the ZnO host matrix.

A similar type of study was performed by Choi et al. [92] for ZnO films implanted with Co (2×10^{16} ions/cm^2) and Fe (1×10^{16} ions/cm^2) by molecular beam epitaxy (MBE) method and then irradiated them with the 200 MeV Ag ion beam. XRD data reveal the occurrence of Co (111) peak at $2\theta = 44.35°$ along with a shoulder peak associated with the CoO phase at $2\theta = 73.5°$ for Co implanted sample. Similarly, the peak corresponding to Fe displays its appearance for Fe implanted. These results reveal that implantation of both the Fe and Co metals leads to cluster formation instead of the occupying Zn site in the host matrix. The in-depth analysis of the peaks associated with these clusters indicates that these clusters are in the nanometre scale. Furthermore, these clusters can be dissolved/removed by employing an SHI beam of 200 MeV Ag irradiation to the fluence of 1×10^{12} ions/cm^2. Temperature-dependent resistivity measurements of un-implanted film display metallic behavior up to 250 K and then it exhibits semiconductor type behavior due to the existence of V_O in the films. Similar variation is displayed by Co implanted samples at a transition temperature (from metallic to semiconducting) of 285 K however resistivity value increased by a factor of 2.5. The enhancement in the resistivity after Co implantation is due to the occurrence of Co clusters which act as scattering centers. However, SHI irradiation reduces the resistivity value by half as well as the disappearance of metal–semiconductor transition. A similar kind of variation is seen for Fe implanted films. SHI irradiation by Ag ion beams dissolves these clusters resulting in the reduction in resistivity values. The magnetic behavior is shown in both samples. However, the value of saturation magnetization with Co implantation is almost 10 times higher in comparison to Fe implanted ZnO thin films which is due to the high magnetic moment per Co ions, and observed results are consistent with previous literature [93].

Apart from numerous studies over swift heavy and low-energy ion beam techniques, Yu et al. [94] inspected that magnetic properties could also be tailored with the help of UV irradiation. In this case, two-step methods are used for the deposition of well-aligned ZnO nanorods over a glass substrate. The first step involves the growth of Gallium-doped ZnO thin films via PLD technique as a seed layer on a glass substrate. After that, chemical bath deposition was employed by immersing the seed layer in zinc acetate aqueous solution at 90 °C for 2 h with or without UV light of 365 nm. XRD measurements indicate that there is an intense improvement in the crystalline growth towards the (002) orientation in ZnO nanorods as one irradiates a precise wavelength of ultraviolet (UV) equivalent to the bandgap of ZnO. Both samples, with and without UV irradiated undoped nanorods, exhibit the anisotropic ferromagnetism, however, the enhanced magnetic character has been obtained for the UV irradiated sample. The term anisotropic ferromagnetism is introduced here because the authors have focused their study to find out whether the ferromagnetism in the present case is analogous to that of the multiferroics (where induction direction is perpendicular to the electric field). Therefore, precise UV irradiation significantly enhances ferromagnetic anisotropy. Furthermore, it has been proposed that the degree of anisotropic ferromagnetism could be tuned by employing different UV wavelengths to vary the electric polarization along with magnetism [94].

The relationship between magnetism and defects: ion irradiation, in contrast with introducing defects of various types e.g. anti-site, interstitial, vacant, etc. to identify the specific type of defect that causes magnetism. In addition to electron spin resonance, scanning tunnelling microscopy is an option to locally determine the magnetic moments and their related defects. These research activities utilize irradiation for defect generation in some materials and also some non-magnetic materials. There is an exchange of disturbance in-network and magnetization. Careful characterization of defects along with ion-irradiation is much helpful for the elucidation of the local magnetic moments and coupling mechanism in semiconductors. There is still a lot of research needed for this.

Botsch et al. [95] in their study on ZnO microwires have demonstrated an all-semiconducting spin filter effect. Proton implantation in the Li-doped ZnO microwires was found to induce ferromagnetism in the ZnO by stabilizing the Zn-vacancies created by the Li doping. With the help of electron-beam lithography, proton implantation was used to design N^+-N–N^+type all-semiconducting defect-induced magnetic/nonmagnetic/defect-induced magnetic homojunction with a potential barrier at the magnetic (N^+)/nonmagnetic (N) junction. Due to such a design and potential barrier, a minority spin-filter effect was demonstrated with spin-filter efficiency greater than 10% at room temperature.

5.3 Magnetism in Magnesium Oxide (MgO)

MgO is well known non-magnetic material, however, it exhibits onset of defect-induced magnetism [96]. This makes this material suitable for investigating defect-related phenomena [97]. Vacancies/defects are responsible for inducing magnetism in this material [98–100]. It is observed that a defect (cation vacancies) concentration of the order of 10^{21} cm^{-3} in the non-magnetic materials having structure analogues to MgO is required to percolate magnetism [101]. Thus, ion implantation and irradiation can provide an alternative tool to modify the behavior of this material. However, there are very few studies available on the implantation and irradiation studies on MgO thin films / single crystals. Li et al. [102] implanted the MgO thin films with different low energy ions (C/N/O) at 70 keV with 2×10^{16} and 2×10^{17} ions/cm^2 at room temperature. All samples with high dose implantation exhibit room-temperature ferromagnetism and the implantation of C/N/O in MgO leads to enhancement in magnetization. Based on the experimental observations and various characterization tools, it is confirmed that ferromagnetic behavior in O implanted samples is associated with the Mg vacancies. On the other hand, implantation of N and C ions played a dominant role in the magnetic behavior compared to Mg vacancies in the host lattice. The localized magnetic moment as the crucial factor of ferromagnetic coupling has also been discussed in this report.

Singh et al. [103] investigated the role of 60 keV Fe and Zn ions implantation (at a dose of 5×10^{16} ions/cm^2) on the surface and local electronic structure modification of MgO thin films. The estimated range of Fe ions (31.8 nm) is large than the one of Zn ions (28.6 nm). It was concluded that implantation weakens the coordination of Mg^{2+} ions in the MgO matrix. These ions result in weak magnetic behaviour in MgO [104]. Another report on the effect of different transition metal ions (Co, Cu, and Ni) having 100 keV energy with five different fluences varying from 1×10^{15} ions/cm^2 to 5×10^{16} ions/cm^2 on MgO thin films (250 nm thick) grown over Si (100) substrate was studied by Kaur et al. [105]. It is found that Mg^{2+} ions maintain their charge state in the host matrix and the valence state is not altered even after the interaction of ions of the host matrix as 2+ is the most stable state for MgO. Furthermore, XAS measurements indicate that implanted ions do not oxidize and there is no formation of metal–oxygen (TM-O) like structure. Ide-Ektessabi et al. [106] focused their study on MgO thin films irradiated with oxygen beams. It is concluded that ion beam irradiation during film growth strongly modifies the crystal orientation of the films. Also, it is evident from the analysis that irradiation with oxygen ions results in oxygen-rich MgO thin films.

The introduction of defects in metal-doped MgO using 120 MeV Au$^+$ irradiation affects the magnetization of this material [107]. In the study, MgO single crystals were doped with Ni and Co impurities and explored the variation in magnetic properties after SHI irradiation. A comparison was made between these results and the results obtained from thin films having a large concentration of trapped defects. As grown bulk single crystals having less concentration of vacancies exhibit perfect paramagnetic curves over the entire range of temperature and magnetic field. It is explored that introduction of defects via any means (by irradiation or during thin film deposition), the paramagnetic curve shifts to the ferromagnetic character.

5.4 Zinc Ferrite (ZnFe$_2$O$_4$)

Ion beam based techniques are considered to be suitable for modifying the characteristics of ferrites. This aspect was also investigated by Sharma et al., in Mg$_{0.95}$Mn$_{0.05}$Fe$_2$O$_4$ nanoparticles irradiated by 100 MeV Si and Ni [108]. It was concluded that the electronic stopping value plays a dominant role in the observed change in magnetization for this system. Thus, numerous ferrite systems were investigated using heavy ion irradiation such as Fe$_3$O$_4$ [109], Mg-Mn ferrite [110–112], Aluminium ferrite [113], and Ni-Cu ferrite [114]. These reports envisage the role of defects in altering the physical properties of ferrites.

Bulk ZnFe$_2$O$_4$ is paramagnetic at room temperature [115]. It exhibits antiferromagnetic ordering having a Neel temperature of 10 K. The magnetic ordering can be tuned by particle size [116, 117], method of synthesis [118], and post-synthesis treatment [119]. Defects/Oxygen vacancies created during the method of synthesis strongly influence the magnetic ordering in this material [120, 121]. In addition to this, when ZnFe$_2$O$_4$ systems are irradiated using ion beams of Kr, Xe, and O ions, magnetic ordering induces in the system [122–124]. The induced magnetic behavior of ZnFe$_2$O$_4$ occurs as a result of defects produced in the system by the heavy-ion irradiations [125].

Apart from defects, the ferrite system also exhibits change in cation (Zn^{2+} and Fe^{3+} ions) occupancies among the tetrahedral and octahedral spinel structure of this material induced by ion irradiation [126]. Under 100 MeV ion irradiation, magnetic behavior depends on the annealing temperature utilized to synthesize ZnFe$_2$O$_4$ nanoparticles [127]. There is no change observed in the samples annealed at 500 °C (ZF500). However, modification in magnetic behavior is observed in the samples that were annealed at 1000 °C (ZF1000). (Fig. 5.7). In the case of ZF1000, blocking temperature is 18 K for pristine, however, it increases to 34 K for fluence of 5×10^{13} ions/cm^2 [128]. This may be due to irradiation-induced changes in magnetic interaction that depend on the crystallite size of these nanoparticles [127]. In the case of irradiation using Ag beam (Fig. 5.8), the change in magnetic behavior is reported which depends both on the crystallite size of pristine ZnFe$_2$O$_4$ and the ion fluence of irradiation [129]. In the case of 120 MeV Ag ions, the formation of latent tracks is observed using high-resolution electron microscopy [130].

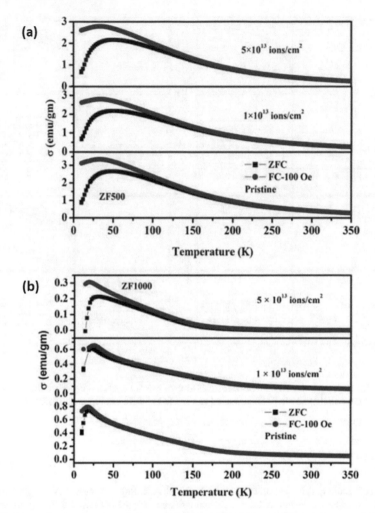

Fig. 5.7 Magnetization for the pristine and irradiated (100 MeV O^{7+}) counterpart of ZnFe$_2$O$_4$ annealed at **a** 500 °C (ZF500) and **b** 1000 °C (ZF1000). ZF 500 and ZF1000 have crystallite sizes of 16 and 62 nm respectively [125]

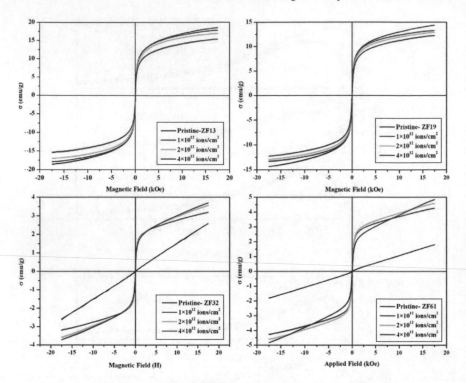

Fig. 5.8 Room temperature magnetic hysteresis loop of the $ZnFe_2O_4$ having crystallite sizes 13 nm (ZF13), 19 nm (ZF19), 32 nm (ZF32), and 61 nm (ZF61) at different fluences of 120 MeV Ag ions [126]

References

1. M.H.F. Sluiter, Y. Kawazoe, P. Sharma, A. Inoue, A.R. Raju, C. Rout, U.V. Waghmare, First principles-based design and experimental evidence for a ZnO-based ferromagnet at room temperature. Phys. Rev. Lett. **187204**, 3–6 (2005)
2. K. Osuch, E.B. Lombardi, W. Gebicki, First-principles study of ferromagnetism in $Ti_{0.0625}Zn_{0.9375}O$. Phys. Rev. B. **73**, 1–5 (2006)
3. L.A. Errico, M. Rentería, Theoretical study of magnetism in transition-metal-doped TiO_2 and $TiO_{2-\delta}$. Phys. Rev. B **67**, 1–8 (2005)
4. R. Janisch, P. Gopal, N.A. Spaldin, Transition metal-doped TiO_2 and ZnO present status of the field. J. Phys.: Condens. Matter. **17**, 657–689 (2005)
5. K. Jindal, M. Tomar, R.S. Katiyar, V. Gupta et al., Structural and magnetic properties of N doped ZnO thin films. J. Appl. Phys. **111**, 102805 (2012)
6. F. Gao, J. Hua, C. Yang, Y. Zheng, H. Qina, L. Suna, X. Kong, M. Jiang, First-principles study of magnetism driven by intrinsic defects in MgO. Solid State Commun. **149**, 855–858 (2009)
7. S. Suman, A. Chahal, P. Kumar, Kumar, Zn doped α-Fe2O3: an efficient material for UV driven photocatalysis and electrical conductivity. Curr. Comput.-Aided Drug Des. **10**, 4 (2020)
8. P.V. Radovanovic, D.R. Gamelin, High-temperature ferromagnetism in Ni^{2+} -doped ZnO aggregates prepared from colloidal diluted magnetic semiconductor quantum dots. Phys. Rev. Lett. **157202**, 1–4 (2003)

9. N.S. Norberg, K.R. Kittilstved, J.E. Amonette, R.K. Kukkadapu, D.A. Schwartz, D.R. Gamelin, Synthesis of colloidal Mn^{2+}: ZnO quantum dots and high-T_C ferromagnetic nanocrystalline thin films. J. Am. Chem. Soc. **14**, 9387–9398 (2004)
10. K. Ueda, H. Tabata, T. Kawai, K. Ueda, H. Tabata T. Kawai, Magnetic and electric properties of transition-metal-doped ZnO films **988**, 67–70 (2012)
11. A. Fujishima, K. Honda, Electrochemical photolysis of water at a semiconductor electrode. Nature **238**, 37–38 (1972)
12. K.M. Glassford, J.R. Chelikowsky, Structural and electronic properties of titanium dioxide. Phys. Rev. B **46**, 1284 (1992)
13. M. Ramamoorthy, R.D. King-Smith, D. Vanderbilt, Defects on TiO_2 (110) surfaces. Phys. Rev. B **49**, 7709 (1994)
14. P.J.D. Lindan, N.M. Harrison, M.J. Gillan, J.A. White, First-principles spin-polarized calculations on the reduced and reconstructed TiO_2 (110) surface. Phys. Rev. B **55**, 15919 (1997)
15. L. Martinu, D. Poitras, Plasma deposition of optical films and coatings: a review. J. Vac. Sci. Technol. A **18**, 2619 (2000)
16. W. Brown, W. Grannemann, C-V characteristics of metal-titanium dioxide-silicon capacitors. Solid-State Electron. **21**, 837–846 (1978)
17. M.R. Hoffmann, S.T. Martin, W. Choi, D.W. Bahnemannt, Environmental applications of semiconductor photocatalysis. Chem. Rev. **95**, 69–96 (1995)
18. V. Kumar, O.M. Ntwaeaborwa, J. Holsa, D.E. Motaung, H.C. Swart, The role of oxygen and titanium related defects on the emission of TiO_2:Tb^{3+} nano-phosphor for blue lighting applications. Opt. Mater. **46**, 510 (2015)
19. X. Chen, L. Liu, P.Y. Yu, S.S. Mao, Increasing solar absorption for photocatalysis with black hydrogenated titanium dioxide nanocrystals. Science **331**, 746–750 (2011)
20. G. Liu, L. Yin, J. Wang, P. Niu, C. Zhen, Y. Xie, H. Cheng, A red anatase TiO_2 photocatalyst for solar energy conversion. Energy Environ. Sci. **5**, 9603–9610 (2012)
21. H. Yu, Y. Zhao, C. Zhou, L. Shang, Y. Peng, Y. Cao, L. Wu, C. Tung, T. Zhang, Carbon quantum dots/TiO_2 composites for efficient photocatalytic hydrogen evolution. J. Mater. Chem. A **2**, 3344–3351 (2014)
22. C. Zhou, Y. Zhao, L. Shang, Y. Cao, L. Wu, C. Tung, T. Zhang, Facile preparation of black Nb^{4+} self-doped $K_4Nb_6O_{17}$ microspheres with high solar absorption and enhanced photocatalytic activity. Chem. Commun. **50**, 9554–9556 (2014)
23. M. Venkatesan, C.B. Fitzgerald, J.M.D. Coey, Unexpected magnetism in a dielectric oxide. Nature **430**, 630 (2004)
24. J. Hu, Z. Zhang, M. Zhao, H. Qin, M. Jiang, Room-temperature ferromagnetism in MgO nanocrystalline powders. Appl. Phys. Lett. **93**, 192503 (2008)
25. N.H. Hong, J. Sakai, N. Poirot, V. Brizé, Room-temperature ferromagnetism observed in undoped semiconducting and insulating oxide thin films. Phys. Rev. B **73**, 132404 (2006)
26. A. Sundaresan, R. Bhargavi, N. Rangarajan, U. Siddesh, C.N.R. Rao, Ferromagnetism as a universal feature of nanoparticles of the otherwise nonmagnetic oxides. Phys. Rev. B **74**, 161306(R) (2006)
27. J.M.D. Coey, M. Venkatesan, P. Stamenov, C.B. Fitzgerald, L.S. Dorneles, Magnetism in hafnium dioxide. Phys. Rev. B **72**, 024450 (2005)
28. H. Peng, H.J. Xiang, S.-H. Wei, S.-S. Li, J.-B. Xia, J. Li, Origin and enhancement of hole-induced ferromagnetism in first-row d^0 semiconductors. Phys. Rev. Lett. **102**, 017201 (2009)
29. F. Wang, Z. Pang, L. Lin, S. Fang, Y. Dai, S. Han, Magnetism in undoped MgO studied by density functional theory. Phys. Rev. B **80**, 144424 (2009)
30. P.D. Esquinazi, W. Hergert, M. Stiller, L. Botsch, H. Ohldag, D. Spemann, M. Hoffmann, W.A. Adeagbo, A. Chassé, S.K. Nayak, H.B. Hamed, Defect induced magnetism in nonmagnetic oxides: basic principles, experimental evidence, and possible devices with ZnO and TiO_2. Phys. Status Solidi B **257**, 1900623 (2020)
31. L. Botsch, P.D. Esquinazi, C. Bundesmann, D. Spemann, Toward a systematic discovery of artificial functional magnetic materials. Phys. Rev. B **104**, 014428 (2021)

32. D.K. Shukla, R. Kumar, S. Mollah, R.J. Choudhary, P. Thakur, S.K. Sharma, N.B. Brookes, M. Knobel, Swift heavy ion irradiation-induced magnetism in magnetically frustrated $BiMn_2O_5$ thin films. Phys. Rev. B **82**, 174432 (2010)

33. A. Benyagoub, Phase transformations in oxides induced by swift heavy ions. Nucl. Instrum. Methods Phys. Res. B **245**, 225 (2006)

34. H. Rath, P. Dash, T. Som, P.V. Satyam, U.P. Singh, P.K. Kulriya, D. Kanjilal, D.K. Avasthi, N.C. Mishra, Structural evolution of TiO_2 nanocrystalline thin films by thermal annealing and swift heavy ion irradiation. J. Appl. Phys. **105**, 074311 (2009)

35. S.P.S. Porto, P.A. Fleury, T.C. Damen, Raman spectra of TiO_2, MgF_2, ZnF_2, FeF_2, and MnF_2. Phys. Rev. **154**, 522 (1967)

36. H. Thakur, R. Kumar, P. Thakur, N.B. Brookes, K.K. Sharma, A.P. Singh, Y. Kumar, S. Gautam, K.H. Chae, Modifications in structural and electronic properties of TiO2 thin films using swift heavy ion irradiation. J. Appl. Phy. **110**, 083718 (2011)

37. H. Thakur, P. Thakur, R. Kumar, N.B. Brookes, K.K. Sharma, A.P. Singh, Y. Kumar, S. Gautam, K.H. Chae, Irradiation induced ferromagnetism at room temperature in TiO_2 thin films: X-ray magnetic circular dichroism characterizations. Appl. Phy. Lett. **98**, 192512 (2011)

38. B. Bharati, N.C. Mishra, C. Rath, Effect of 500 keV Ar^{2+} ion irradiation on structural and magnetic properties of TiO_2 thin films annealed at 900 °C. Appl. Surf. Sci. **455**, 717–723 (2018)

39. P. Mohanty, V.P. Singh, N.C. Mishra, S. Ojha, D. Kanjilal, C. Rath, Evolution of structural and magnetic properties of Co-doped TiO_2 thin films irradiated with 100 MeV Ag^{7+} ions. J. Phys. D: Appl. Phys. **47**, 315001 (2014)

40. D. Sanyal, M. Chakrabarti, P. Nath, A. Sarkar, D. Bhowmick, A. Chakrabarti, Room temperature ferromagnetic ordering in 4 MeV Ar^{5+} irradiated TiO_2. J. Phys. D: Appl. Phys. **47**, 025001 (2010)

41. J.F. Ziegler, J.P. Biersack, in *The Stopping and Range of Ions in Matter* (New York, Pergamon, 1985), pp. 93–129

42. D.R. Locker, J.M. Meese, Displacement thresholds in ZnO. IEEE Trans. Nucl. Sci. **19**, 237 (1972)

43. N. Ishikawa, S. Yamamoto, Y. Chimi, Structural changes in anatase TiO_2 thin films irradiated with high-energy heavy ions. Nucl. Instrum. Methods B **250**, 250 (2006)

44. L.R. Shah, H. Zhu, W.G. Wang, B. Ali, T. Zhu, X. Fan, Y.Q. Song, Q.Y. Wen, H.W. Zheng, S.I. Shah, Effect of Zn interstitials on the magnetic and transport properties of bulk Co-doped ZnO. J. Phys. D: Appl. Phys. **43**, 035002 (2010)

45. S. Zhou, C.E. Izmar, K. Potzger, M. Krause, G. Talut, M. Helm, J. Fassbender, S. Zvyagin, J. Wosnitza, H. Schmidt, Origin of magnetic moments in defective TiO_2 single crystals. Phys. Rev. B. **79**, 113201 (2009)

46. M. Robinson, N.A. Marks, G.R. Lumpkin, Structural dependence of threshold displacement energies in rutile, anatase and brookite TiO_2. Mater. Chem. Phys. **147**, 311–318 (2014)

47. A.K. Rumaiz, B. Ali, A. Ceylan, M. Boggs, T. Beebe, S. Ishmat Shah, Experimental studies on vacancy induced ferromagnetism in undoped TiO_2. Sol. State Commun. **144**, 334–338 (2007)

48. X. Wei, R. Skomski, B. Balamurugan, Z.G. Sun, Stephen Ducharme, D.J. Sellmyer, Magnetism of TiO and TiO_2 nanoclusters. J Appl. Phys. **105**, 07C517 (2009)

49. D. Kim, J. Hong, Y.R. Park, K.J. Kim, The origin of oxygen vacancy induced ferromagnetism in undoped TiO_2. J. Phys. Condens. Matter. **21**, 195405 (2009)

50. O. Vázquez-Robaina, A.F. Cabrera, A.F. Cruz, C.E.R. Torres, Observation of room temperature ferromagnetism induced by high-pressure hydrogenation of anatase TiO_2. The J. Phys. Chem. C **125**, 14366–14377 (2021)

51. H. Dai, X. Li, X. Cai, R. Wei, The magnetism of titanium defected undoped rutile TiO_2: first-principles calculations. Phys. Chem. Chem. Phys. **22**, 25930–25935 (2020)

52. H.H. Huang, P.L. Chan, Effects of MnO_2 doping in V_2O_5-doped ZnO varistor system. M. Chem. and Phy. **75**, 61–66 (2002)

53. J.C. Johnson, H. Yan, P. Yang, R.J. Saykally, Optical cavity effects in ZnO nanowire lasers and waveguides. J. Phys. Chem. B **107**, 8816–8828 (2003)
54. N.G. Romanov, D.O. Tolmachev, A.G. Badalyan, R.A. Babunts, P.G. Baranov, V.V. Dyakonov, Spin-dependent recombination of defects in bulk ZnO crystals and ZnO nanocrystals as studied by optically detected magnetic resonance. Phys. B Phys. Condens. Matter **404**, 4783–4786 (2009)
55. H.T. Ng, J. Han, T. Yamada, P. Nguyen, Y.P. Chen, M. Meyyappan, Single crystal nanowire vertical surround-gate field-effect transistor. Nano Lett. **4**, 1–6 (2004)
56. Z. Fan, J.G. Lu, Z. Fan, J.G. Lu, Gate-refreshable nanowire chemical sensors Gate-refreshable nanowire chemical sensors. Appl. Phys. Lett. **86**, 2003–2006 (2005)
57. S.K. Gupta, A. Joshi, M. Kaur, Development of gas sensors using ZnO nanostructures. J. Chem. Sci. **122**, 57–62 (2010)
58. H. Cheng, C. Chen, C. Tsay, C. Chen, Transparent ZnO thin-film transistor fabricated by sol-gel and chemical bath deposition combination method. Appl. Phys. Lett. **90**, 012113 (2007)
59. L. Guo, Y.L. Ji, H. Xu, P. Simon, Regularly shaped, single-crystalline ZnO nanorods with wurtzite structure. J. Am. Chem. Soc. **124**, 14864–14865 (2002)
60. A. Tabib, W. Bouslama, B. Sieber, A. Addad, H. Elhouichet, M. Ferid, R. Boukherroud, Structural and optical properties of Na doped ZnO nanocrystals: application to solar photocatalysis. Appl. Surf. Sci. **396**, 1528–1538 (2016)
61. P. Gopal, N.A. Spaldin, Magnetic interactions in transition-metal-doped ZnO: An ab initio study. Phys. Rev. B **74**, 1–9 (2006)
62. L. Wei, L. Zhang, Y. Zhang, W.F. Zhang, Enhanced ultraviolet photoluminescence from V-doped ZnO thin films prepared by a sol-gel process. Phys. Status Solidi (a) **204**, 2426–2430 (2007)
63. V. Kumar, O.M. Ntwaeaborwa, T. Soga, V. Dutta, H.C. Swart, Rare earth doped zinc oxide nanophosphor powder: a future material for solid-state lighting and solar cell. ACS Photonics **4**, 2613–2637 (2017)
64. J.A. Phys, K. Kumar, A contrast in the electronic structures of B ion-implanted ZnO thin films grown on glass and silicon substrates by using x-ray absorption spectroscopy. J. Appl. Phys. **128**, 065303 (2020)
65. D. Gao, Z. Zhang, J. Fu, Y. Xu, J. Qi, D. Xue, Room-temperature ferromagnetism of pure ZnO nanoparticles. J. Appl. Phys. **105**, 113928 (2009)
66. S. Majumder, D. Paramanik, A. Gupta, S. Varma, Observation of magnetic domains in undoped ZnO grains at room temperature. App. Surf. Sci. **256**, 513–516 (2009)
67. Y. Li, R. Deng, B. Yao, G. Xing, D. Wang, Tuning ferromagnetism in $Mg_xZn_{1-x}O$ thin films by band gap and defect engineering. Appl. Phys. Lett. **97**, 102506 (2010)
68. A. Sundaresan, R. Bhargavi, N. Rangarajan, U. Siddesh, C.N.R. Rao, Ferromagnetism as a universal feature of nanoparticles of the otherwise nonmagnetic oxides. Phys. Rev. B **74**, 1–4 (2006)
69. D. Sanyal, M. Chakrabarti, T.K. Roy, A. Chakrabarti, The origin of ferromagnetism and defect-magnetization correlation in nanocrystalline ZnO. Phy. Lett. A **371**, 482–485 (2007)
70. T.S. Herng, S.P. Lau, S.F. Yu, J.S. Chen, K.S. Teng, Zn-interstitial-enhanced ferromagnetism in Cu-doped ZnO films. J. Magn. Magn. Mater. **315**, 107–110 (2007)
71. Y. Zhang, E. Xie, Nature of room-temperature ferromagnetism from undoped ZnO nanoparticles. J. Appl. Phys. **99**, 955–960 (2010)
72. A.L. Schoenhalz, J.T. Arantes, A. Fazzio, G.M. Dalpian, Surface magnetization in non-doped ZnO nanostructures. Appl. Phys. Lett. **162503**, 2007–2010 (2009)
73. P. Kumar, V. Sharma, A. Sarwa, A. Kumar, Surbhi, R. Goyal, K. Sachdeva, S. Annapoorni, K. Asokan, D. Kanjilal, Understanding the origin of ferromagnetism in Er-doped ZnO system. RSC Adv. **6**, 89242–89249 (2016)
74. D. Kim, J. Yang, J. Hong, Ferromagnetism induced by Zn vacancy defect and lattice distortion in ZnO. J. Appl. Phys. **106**, 013908 (2009)

75. K. Rainey, J. Chess, J. Eixenberger, D.A. Tenne, C.B. Hanna, A. Punnoose, Defect induced ferromagnetism in undoped ZnO nanoparticles. J. Appl. Phys. **115**, 17D727 (2014)
76. R. Bhardwaj, B. Kaura, J.P. Singh, M. Kumar, H.H. Lee, P. Kumar, R.C. Meena, K. Asokan, K.H. Chae, N. Goyal, S. Gautam, Role of low energy transition metal ions in interface formation in ZnO thin films and their effect on magnetic properties for spintronic applications, appl. Surf. Sci. **479**, 1021–1028 (2019)
77. I. Lorite, B. Straube, H. Ohldag, P. Kumar, M. Villafuerte, P. Esquinazi, C.E. Rodr´ıguez Torres, S.P. Heluani, V.N. Antonov, L.V. Bekenov, A. Ernst, M. Hoffmann, S.K. Nayak, W.A. Adeagbo, G. Fischer, W. Hergert, Advances in methods to obtain and characterise room temperature magnetic ZnO. Appl. Phys. Lett. **106**, 082406, IF. 3.791 (2015)
78. P. Srivastava, S. Ghosh, B. Joshi, P. Satyarthi, P. Kumar, D. Kanjilal, D. Buerger, S. Zhou, H. Schmidt, A. Rogalev, F. Wilhelm, J. Appl. Phys. **111**, 013715 (2012)
79. P. Satyarthi, S. Ghosh, B. Pandey, P. Kumar, C.L. Chen, C.L. Dong, W.F. Pong, D. Kanjilal, K. Asokan, P. Srivastava, J. Appl. Phys. **113**, 183708 (2013)
80. P. Satyarthi, S. Ghosh, P. Mishra, B.R. Sekhar, F. Singh, P. Kumar, D. Kanjilal, R.S. Dhaka, P. Srivastava, J. Magn. Magn. Mater. **385**, 318–325 (2015)
81. P. Kumar, H.K. Malik, S. Gautam, K.H. Chae, K. Asokan, D. Kanjilal, Modifications in room temperature ferromagnetism by dense electronic excitations in $Zn_{0.9}Mg_{0.1}O$ thin films. J. Alloys Compounds **710**, 831–835 (2017)
82. G. Jayalakshmi, K. Saravanan, S. Balakumar, T. Balasubramanian, Swift heavy ion induced modifications in structural, optical and magnetic properties of pure and V doped ZnO films. Vacuum **95**, 66–70 (2013)
83. R. Kumar, F. Singh, Single phase formation of Co-implanted ZnO thin films by swift heavy ion irradiation :optical studies. J. Appl. Phys. **100**, 113708 (2006)
84. S. Guner, O. Gurbuz, S. Caliskan, V.I. Nuzhdin, R. Khaibullin, M. Ozturk, N. Akdogan, The structural and magnetic properties of Co+ implanted ZnO films, app. Surf. Scie. **310**, 235–241 (2014)
85. Z.L. Lu, H.S. Hsu, Y.H. Tzeng, F.M. Zhang, Y.W. Du, J.C.A. Huang, The origins of ferromagnetism in Co-doped ZnO single crystalline films: from bound magnetic polaron to free carrier-mediated exchange interaction. Appl. Phys. Lett. **95**, 102501 (2009)
86. B. Pandey, S. Ghosh, P. Srivastava, P. Kumar, D. Kanjlal, Influence of microstructure on room temperature ferromagnetism in Ni implanted nanodimensional ZnO films. J. Appl. Phys. **105**, 033909 (2009)
87. D.D. Wang, B. Zhao, N. Qi, Z.Q. Chen, A. Kawasuso, Vacancy-mediated ferromagnetism in Co-implanted ZnO studied using a slow positron beam. J. Mater. Sci. **52**, 7067–7076 (2017)
88. T.S. Herng, S.P. Lau, L. Wang, B.C. Zhao, S.F. Yu, M. Tanemura, A. Akaike, K.S. Teng, Magnetotransport properties of p-type carbon-doped ZnO thin films. Appl. Phys. Lett. **95**, 012505 (2009)
89. T.S. Herng, S.P. Lau, C.S. Wei, L. Wang, B.C. Zhao, M. Tanemura, Y. Akaike, Stable ferromagnetism in p-type carbon-doped ZnO nanoneedles. Appl. Phys. Lett. **95**, 133103 (2009)
90. X. Wang, X. Chen, R. Dong, Y. Huang, W. Lu, Ferromagnetism in carbon-doped ZnO films from the first-principle study. Phys. Lett. A **373**, 3091–3096 (2009)
91. M.W. Khan, R. Kumar, M.A.M. Khan, B. Angadi, Y.S. Jung, W.K. Choi, J.P. Srivastva, Solubility of Co clusters in Co-implanted ZnO thin films by 200 MeV Ag^{15+} ions irradiation. Semicond. Sci. Technol. **24**, 095011 (2009)
92. W.K. Choi, B. Angadi, H.C. Park, J.H. Lee, J.H. Song, R. Kumar, Ferromagnetic property of Co and Fe-implanted ZnO thin film at room temperature. Mat. Sci. and Engg. **52**, 42–47 (2006)
93. M. Venkatesan, C.B. Fitzgerald, J.G. Lunney, J.M.D. Coey, Anisotropic ferromagnetism in substituted zinc oxide. Phys. Rev. Lett. **93**, 177206 (2004)
94. C.F. Yu, S.J. Sun, H.S. Hsu, UV irradiations enhance anisotropy of ZnO nanorods in crystal growth and ferromagnetism. Phy. Lette. A **379**, 211–213 (2015)

95. L. Botsch, I. Lorite, Y. Kumar, P.D. Esquinazi, J. Zajadacz, K. Zimmer, A.C.S. Appl, Electron. Mater. **1**, 1832–1841 (2019)
96. J.P. Singh, K.H. Chae, d° ferromagnetism of magnesium oxide. Condensed Matter **2**, 36 (2017)
97. J.P. Singh, W.C. Lim, K.H. Chae, An interplay among the Mg^{2+} ion coordination, structural order, oxygen vacancies and magnetism of MgO thin films. J. Alloys Compd. **806**, 1348–1356 (2019)
98. J.P. Singh, C.L. Chen, C.L. Dong, J. Prakash, D. Kabiraj, D. Kanjilal, W.F. Pong, K. Asokan, Role of surface and subsurface defects in MgO thin film: XANES and magnetic investigations. Superlattices Microstruct. **77**, 313–324 (2015)
99. N. Rani, S. Chahal, P. Kumar, A. Kumar, R. Shukla, S.K. Singh, MgO nanostructures at different annealing temperatures for d0 ferromagnetism. Vacuum **179**, 109539 (2020)
100. B. Choudhury, U. Saikia, M.B. Sahariah, A. Choudhury, Vacancy induced p-orbital ferromagnetism in MgO nanocrystallite. J. Alloys Compd. **819**, 153060 (2020)
101. J. Osorio-Guillén, S. Lany, S.V. Barabash, A. Zunger, Magnetism without magnetic ions: percolation, exchange, and formation energies of magnetism-promoting intrinsic defects in CaO. Phys. Rev. Lett. **96**, 107203 (2006)
102. Q. Li, B. Ye, Y. Hao, J. Liu, J. Zhang, L. Zhang, W. Kong, H. Weng, B. Ye, Room-temperature ferromagnetism observed in C-/N-/O-implanted MgO single crystals. Chem. Phys. Lett. **556**, 237–241 (2012)
103. J.P. Singh, W.C. Lim, J. Lee, J. Song, I.-J. Lee, K.H. Chae, Surface and local electronic structure modification of MgO film using Zn and Fe ion implantation. Appl. Surf. Sci. **432**, 132–139 (2018)
104. J.P. Singh, W.C. Lim, J. Song, S. Lee, K.H. Chae, Fe^+ and Zn^+ ion implantation in MgO single crystals. Mat. Lett. **301**, 130232 (2021)
105. B. Kaur, R. Bhardwaj, J.P. Singh, K. Asokan, K.H. Chae, N. Goyal, S. Gautam, Valence state and co-ordination of implanted ions in MgO. AIP Conf. Proc. **2220**, 090003 (2020)
106. A. Ide-Ektessabi, H. Nomura, N. Yasui, Y. Tsukuda, Ion beam processing of MgO thin films. Thin Solid Films **447–448**, 383–387 (2004)
107. J. Narayan, S. Nori, D.K. Pandya, D.K. Avasthi, A.I. Smirnov, Defect dependent ferromagnetism in MgO doped with Ni and Co. Appl. Phys. Lett. **93**, 082507 (2008)
108. S.K. Sharma, R. Kumar, V.V. Sivakumar, M. Knobel, V.R. Reddy, A. Gupta, M. Singh, Role of electronic loss on the magnetic properties of $Mg0.95Mn0.05Fe_2O_4$. Nucl. Instr. Meth. Phys. Res. B **248**, 37–41 (2006)
109. A. Fnidiki, F. Studer, J. Teillet, J. Juraszek, H. Pascard, S. Meillon, Damage processes in Fe_3O_4 magnetic insulator irradiated by swift heavy ions. Experimental results and modelization. Eur. Phys. J. B **24**, 291–295 (2001)
110. M. Singh, A. Dogra, R. Kumar, Effect of 50 MeVLi3+ ion irradiation on structural, dielectric, and permeability studies of In3+ substituted Mg-Mn ferrite. Nucl. Instr. Methods Phys. Res. B **196**, 315–323 (2002)
111. S. Ghosh, P. Ayyub, N. Kumar, S.A. Khan, D. Banerjee, In situ monitoring of electrical resistance of nanoferrite thin film irradiated by 190 MeV Au14+ ions. Nucl. Instr. Methods Phys. Res. B **212**, 510–515 (2003)
112. S. Ghosh, A. Gupta, P. Ayyub, N. Kumar, S.A. Khan, D. Banerjee, R. Bhattacharya, Swift heavy ion irradiation-induced damage creation in nanocrystalline Li–Mg ferrite thin films. Nucl. Instr. Methods Phys. Res. B **225**, 310–317 (2004)
113. M.C. Chhantbar, K.B. Modi, G.J. Baldha, H.H. Joshi, R.V. Upadhyay, R. Kumar, Influence of 50 MeV Li^{3+} ion irradiation on structural and magnetic properties of Ti4+ substituted $Li_{0.5}Al_{0.1}Fe_{2.4}O_4$. Nucl. Instr. Methods Phys. Res. B. **244**, 124–127 (2006)
114. S.N. Dolia, R. Kumar, S.K. Sharma, M.P. Sharma, S. Chander, M. Singh, Magnetic behavior of nanocrystalline Ni-Cu ferrite and the effect of irradiation by 100 MeV Ni ions. Curr. Appl. Phys. **8**, 620–625 (2008)
115. G.A. Pettit, D.W. Forester, Mössbauer study of cobalt-zinc ferrites. Phys. Rev. B **4**, 3912 (1971)

116. J.P. Singh, R.C. Srivastava, H.M. Agrawal, R.P.S. Kushwaha, P. Chand, R. Kumar, EPR study of nanostructured zinc ferrite. Int. J. Nanosci **7**, 21–27 (2008)

117. J.P. Singh, G. Dixit, R.C. Srivastava, H.M. Agrawal, V.R. Reddy, A. Gupta, Observation of bulk-like magnetic ordering below the blocking temperature in nanosized zinc ferrite. J. Magn. Magn. Mater. **324**, 2553–2559 (2016)

118. C. Yao, Q. Zeng, G.F. Goya, T. Torres, J. Liu, H. Wu, M. Ge, Y. Zeng, Y. Wang, J.Z. Jiang, $ZnFe_2O_4$ nanocrystals: synthesis and magnetic properties. J. Phys. Chem. C **111**, 12274–12278 (2007)

119. E.V. Gafton, G. Bulai, O.F. Caltun, S. Cerver, S. Macé, M. Trassinelli, S. Steydli, D. Vernhet, Structural and magnetic properties of zinc ferrite thin films irradiated by 90 keV neon ions. Appl. Surf. Sci. **379**, 171–178 (2016)

120. C.E. Rodríguez Torres, F. Golmar, M. Ziese, P. Esquinazi, S.P. Heluani, Evidence of defect-induced ferromagnetism in $ZnFe_2O_4$ thin films. Phys. Rev. B **84**, 064404 (2011)

121. C.E. Rodríguez Torres, G.A. Pasquevich, P. Mendoza Zélis, F. Golmar, S.P. Heluani, S.K. Nayak, W.A. Adeagbo, W. Hergert, M. Hoffmann, A. Ernst, P. Esquinazi, S.J. Stewart, Oxygen-vacancy-induced local ferromagnetism as a driving mechanism in enhancing the magnetic response of ferrites. Phys. Rev. B **89**, 104411 (2014)

122. F. Studer, C. Houpert, D. Groult, J.Y. Fan, A. Meftan, M. Toulemonde, Spontaneous magnetization induced in the spinel $ZnFe_2O_4$ by heavy ion irradiation in the electronic stopping power regime. Nucl. Instr. Methods Phys. Res. B **82**, 91–102 (1993)

123. C. Upadhyay, H.C. Verma, C. Rath, N.C. Misra, N.S.C. Annual Report 2001–02, pp. 135

124. M. Satalkar, S.N. Kane, S. Raghuvanshi, On the $16O^{6+}$ ion irradiation induced magnetic moment generation in $ZnFe_2O_4$ nano ferrite. AIP Conf. Proc. 1953, 030069 (2018)

125. R.C. Srivastava, J.P. Singh, H.M. Agrawal, R. Kumar, A. Tripathi, R.P. Tripathi, V.R. Reddy, A. Gupta, 57Fe Mössbauer investigation of nanostructured zinc ferrite irradiated by 100 MeV oxygen beam. J. Phys. Conf. Ser. **217**, 012109 (2010)

126. J.P. Singh, R.C. Srivastava, H.M. Agrawal, R. Kumar, V.R. Reddy, A. Gupta, Magnetic study of nanostructured zinc ferrite irradiated with 100 MeV O-beam. J. Magn. Magn. Mater. **322**, 1701–1705 (2010)

127. J.P. Singh, R.C. Srivastava, H.M. Agrawal, R. Kumar, Magnetic behavior of nanosized zinc ferrite under heavy ion irradiation. Nucl. Instrum. Methods Phys. B **268**, 1422–1426 (2010)

128. J.P. Singh, R.C. Srivastava, H.M. Agrawal, P. Chand, R. Kumar, Observation of size-dependent attributes on the magnetic resonance of irradiated zinc ferrite nanoparticles. Curr. Appl. Phys. **11**, 532–537 (2011)

129. J.P. Singh, G. Dixit, R.C. Srivastava, H. Kumar, H.M. Agrawal, R. Kumar, Study of size-dependent features of swift heavy ion irradiation in nanosized zinc ferrite. J. Magn. Magn. Mater. **324**, 3306–3312 (2012)

130. J.P. Singh, G. Dixit, H. Kumar, R.C. Srivastava, H.M. Agrawal, R. Kumar, Formation of latent tracks and their effects on the magnetic properties of nanosized zinc ferrite. J. Magn. Magn. Mater. **352**, 36–44 (2014)

Chapter 6
Summary and Future Prospective

In this book, detailed information about ion beam interactions and ion beam-induced defects in oxide-based materials such as MgO, ZnO, TiO_2, and $ZnFe_2O_4$ were presented and explored their potential applications. The summary and future perspective of the ion beam-induced modifications in oxide semiconductors are discussed in detail in the subsequent sections.

6.1 Summary

The work presented in this book has shown that the energetic ion beams (both at high energy and low energies) can be exploited in different ways such as defects creations/annihilations, and modification of the physical properties with emphasis on magnetism. This effect on the materials depend upon the energy and fluence of ions and the type of ion species. The interaction of ions with material is the deciding factor in ion beam-induced modifications. Several experiments provide a detailed approach into the ion beam mechanism and its reaction with the target, new possibilities in sample characterization, analysis, development of materials, modification of the selective properties, etc., allowing us to make a moderately direct evaluation with structural calculations. The ion beam process offers a well controllable set of tools that can be used for the scientific survey at the atomic level and eventually controlled and exploited at the macroscopic level. In this brief overview, we have provided an outline of the improvements, which have taken place in elementary and functional characteristics of ion–solid interactions. We have also briefly described ion beam accelerators, including ion beam terminology based on their type of interaction and applications of numerous kinds of such interactions in current commercial technology. Besides, diffraction and microscopy of oxide materials, including observations of the irradiation and implantation prospect allocations at the surfaces and interfaces of oxide semiconductor thin film for magnetic application, which are

© The Author(s), under exclusive license to Springer Nature Switzerland AG 2022   59
P. Kumar et al., *Ion Beam Induced Defects and Their Effects in Oxide Materials*,
SpringerBriefs in Physics,
https://doi.org/10.1007/978-3-030-93862-8_6

strongly connected to the episodic prospective of ion beam modifications. In solid-state physics, there is a challenge on the dependence of electrical conductivity on selected doping atoms and associated defects. This needs to be understood because it affects the devices.

It is discussed in detail that various effects, depends on the ion's energy and the nature of the target structure, in terms of low energy irradiation (ion implantation) at surfaces along with suppressed modification in case of doping. The nuclear particle track and modifications of oxide materials have found enormous possibilities in the domain of microelectronics. The dopants can be implanted for modifications of electronic conduction at places of interest by suitable selection of ion species, energy, and dose. The SHI has proven to produce diverse defects; partially annealed the defects that are initially present in the material, created phase transformation, amorphization or crystallization, anisotropic growth of the material, microstructure modification, etc. It has been found that the effects depend not only on the magnitude of energy deposited in the target material but it also depends on the fluence. The microscopic understanding is envisaged in terms of existing models, mainly the Coulomb explosion, thermal spike, model. A suitable theoretical model to recognize a variety of observations is yet to evolve. The details of defect formation and surface modifications are explained by using the examples of TiO_2, ZnO, MgO and $ZnFe_2O_4$. There is a rich variety of phenomena observed in these systems from the physics point of view, which needs further investigation. One such phenomenon is the so-called defect induced magnetism. According to the prevalent knowledge of magnetism, systems with localized moments are essential for a system to be ferromagnetic. However, systems with no ions/atoms having localized moments, show ferromagnetism like pure ZnO, TiO_2 and various other oxide materials. The origin of this magnetism is generally attributed to the defects present in the structure. There are studies on these issues but the field is still new and hence a fertile ground for the researchers to explore new horizons in ion-induced physics in oxides. A detailed experimental and theoretical work is required to check the various sources of magnetism. As an experimentalist, we need to check the various theories proposed for explaining the results and take initiative to design new experiments, which can help in unravelling the true nature of magnetism in oxide materials.

6.2 Future Prospective

There is still a need to have more active and dynamic studies related to the ion beam-induced defects in oxide materials focusing on their physical properties like optical and magnetism. Since the creation of defects in oxide materials are transient processes, probing of these phenomena need an *in–situ* monitoring of the transient response of the system during implantation and irradiation. To understand the defects creation and ion-induced modification, there is a need to study further by *in-situ* characterization tools such as proton-induced X-ray emission (PIXE) for elemental analysis, synchrotron-based X-ray diffraction for detailed phase analysis, measurement of near and far edges using X-ray absorption spectroscopy study for bonding and charge state, synchrotron X-ray photoelectron spectroscopy (binding energy, charge state) and X-ray magnetic circular dichroism (XMCD) for detailed magnetism, etc. These techniques provide an in-depth understanding of the various physical phenomena that might be responsible for the creation/annihilation of defects which in turn are responsible for the magnetism in these oxide semiconductors. A discussion on these techniques is beyond the scope of our book. Researchers are putting efforts to understand the phenomena with the help of synchrotron-based spectroscopic techniques.

Printed in the United States
by Baker & Taylor Publisher Services